W9-APY-656

Americans
and
Environment

THE McNEIL CENTER FOR
EARLY AMERICAN STUDIES

DEACCESSIONED

PROBLEMS IN
AMERICAN CIVILIZATION

Under the editorial direction of
Edwin C. Rozwenc

Americans and Environment

The Controversy over Ecology

Edited and with an Introduction by

John Opie
Duquesne University

D. C. HEATH AND COMPANY
Lexington, Massachusetts Toronto London

Copyright © 1971 by D. C. Heath and Company.

All rights reserved. No part of this publication may be reproduced or transmitted in any form or by any means, electronic or mechanical, including photocopy, recording, or any information storage or retrieval system, without permission in writing from the publisher.

Published simultaneously in Canada.

Printed in the United States of America.

International Standard Book Number: 0–669–61614–1

Library of Congress Catalog Card Number: 76-160030

CONTENTS

III CONTEMPORARY ENVIRONMENTAL PROBLEMS

IV "ON A CLEAR DAY YOU CAN SEE FOREVER"—ALTERNATIVE SOLUTIONS

INTRODUCTION

Interest in the well-being of the environment is an unexpected development in American history. When John Winthrop listed reasons in 1629 for migration to the New World, he reaffirmed a long-standing Western belief that man's own survival depended upon a rough-hewn mastery over nature. This utilitarian claim had divine sanction: "The whole earth is the Lord's garden, and he hath given it to the sons of men," and with a general condition, Genesis 1:28: "Increase and multiply, replenish the earth and subdue it." Andrew Jackson, in his 1830 inaugural address, believed that the raw wilderness was the major obstacle to America's material and spiritual progress: "What good man would prefer a country covered with forests and ranged by a few thousand savages to our extensive Republic, studded with cities, towns, and prosperous farms, embellished with all the improvements which art can devise or industry execute?" In the last half of the twentieth century, the well-being of American civilization is still defined by its economic growth, industrial progress, and capital development.

The ecological controversy broke out between 1962 and 1970 when many Americans concluded that their accustomed growth, progress, and development also meant they were beginning to live in an unhealthy and deteriorating environment. America's air and water had become polluted as to be increasingly unusable. Her natural resources were rapidly disappearing, without effective replenishment or significant energy replacement. Population growth and urban crowding have led Americans to complain that they have unwittingly lost an expansive sense of freedom which they once believed was their birthright. Life, liberty, and the pursuit of happiness seemed to slip from their grasp as Americans were being exposed to more noise, traffic, crowding, ugliness, and pollution. They are surprised by the narrowing spectrum of liberties once taken for granted. "America the Beautiful" had become "America the Befouled." A California group called Ecology Action rewrote the song's lyrics:

> *O, cancerous for smoggy skies, for pesticided grain.*
> *Irradiated mountains rise about an asphalt plain.*
> *America, America, thy birds have fled from thee.*
> *Thy fish lie dead by poisoned streams from sea to fetid sea.*
>
> *America, America, thy sins prepare thy doom:*
> *Monoxide clouds shall be thy shroud, thy cities be thy tomb.*

American industrialists, politicians, and consumers are shocked when a new breed of scientists—the ecologists—call them predators, destroying angels, or outlaws because they misuse enormous power that controls and changes the condition of the earth. Ironically, the American quest for a better way of life may mean a poorer way of life because the environment is weakened, perhaps fatally. One writer has concluded that "it may be that this caterwauling creature, man, has at last goaded the cosmos into stepping on him."

Ecology is the study of living things in relation to their environment and to each other. The term was coined a century ago from two Greek words meaning "the study of the home." Ecologists argue that the course of American history included mounting disregard for man's biological foundations as the nation moved out of a wilderness and agricultural condition into the "artificial environment" fabricated out of technology and industry. They speak of an indispensable "world we have lost." This is usually not a romantic back-to-nature primitivism, but a scientifically oriented biological, psychological, and sociological concern. They argue that man's biological nature must again receive proper attention, as man is an organism whose home in the natural world can never be discarded, even for the sake of "higher things" of the mind or spirit. Americans still need air to breathe, water to drink, and suitable food. Peter Farb has said that man's unprecedented mastery of fire, shelter, and clothing still leaves unanswered the question "whether [they] will be used to maintain the delicate ecological relationships that have slowly developed over these past million or so years of life—or whether they will be used to rip apart the elaborate tapestry of life—man himself must determine." There is an evangelistic and prophetic zeal in many of their pronouncements: Robert Reinow and Leona Train Reinow say,

> *It is with sadness that we look out today upon a tired and beat-up countryside, a land with scrub woodlands, a marching inner desert, shrinking natural resources and sinking water tables, a land fouled with man's wastes and pungently interlaced with the world's longest and most notorious sewers. Our industrial fervor has outdistanced our social values —even our common sense. And, underlying it all, our breeding fervor has outdistanced all concern for the future.*

Ecologists are dismayed when historians justify their teaching by saying that "man has no biology any longer, just history." American history is seen in a different light when ecologists claim that the American continent was better off when it had no history, just biology. In the absence of modern civilized man, all living things were united in a delicate interdependence which sustained life in ways that have never been recovered. The ecological perspective could revise our views of the frontier, manifest destiny, even our definitions of freedom and poverty. Certainly two traditional elements of American civilization—expansionism and capitalism—could be extensively retouched in the light of the ecological controversy. The debate is still difficult, however, to position in American history. It has become more than a question of individual issues, like wilderness conservation, pollution control, or preservation of energy resources. All of these are included, but in its broadest perspective, ecology is suggestive of a highly integrative style of life which includes man's social and intellectual experience as well as his biological existence. It is an attempt to remake American life.

The opening selection from Ralph Waldo Emerson depicts a nineteenth-century romantic affirmation of the natural world as a source of philosophical values. American transcendentalists sought harmony—even some kind of mystical union—between nature, man,

and God. In the second selection, George Perkins Marsh is represented as an early interpreter of the earth as a limited resource, and
yet immensely supportive for human development. The third selection
is a bully appreciation of the revitalizing powers of the American
wilderness by Theodore Roosevelt, who led an American conservationist crusade. The fourth and fifth selections are by John Muir
and Gifford Pinchot, at first close associates in the environmental
crusade. In the selections, however, they reflect growing partisan-
ship as they fought bitterly over whether the environment ought
to be preserved as sheer wilderness, or the wild rivers and primeval
forests be managed to help build a civilization. The twentieth-
century naturalist Aldo Leopold is the founding father of today's
broader humanistic ecological concern. In the selection he argues
for a new extended ethics which would include ties and obligations
to the natural world.

The modern controversy over man's use of the earth broke out in
1960 when Rachel Carson first published *Silent Spring,* her eloquent
essay on the irreversible damage that pesticides have done to the
balance of nature. The public furor that followed is detailed by a
selection from Frank Graham's *Beyond Silent Spring,* written from
the perspective of 1970. In the late 1960's and early 1970's, specific
environmental problems gained widespread attention. At this later
date, limited and immediate problems were examined in the framework of a comprehensive ecological concern which made them
examples of a deeper malais in American society. Paul Ehrlich
insists, in the selection on population, that man cannot hide any
longer from the threat of too many people. The entire movement
began to acquire an apocalyptic flavor and spoke of immediate
crisis. John Burchard describes the poverty and humiliation of life
in unplanned and crowded cities, but he also expresses hope in the
nobility of man and his creative powers. Robert and Leona Reinow
write of a miserable death of the human race as man's own effluvia
make the environment uninhabitable. James Ramsey is frightened
about the disappearance of wilderness just at the point in American
history when mankind needs to recover its mental health by recon-
ciliation with wild nature.

The next several selections reflect upon contemporary environmental questions from several very different perspectives. They all

acknowledge the ecological crisis in American society, and they all admit that future possibilities rest with man's handling of his technology, but they differ widely over the best possible course for the future. The biologist Paul Sears is represented by his pathfinding plea to rearrange American values according to ecology as a "subversive science." He hopes that Americans might recover a sense of man's inescapable biological nature. Loren Eiseley, probably the most eloquent modern naturalist, attempts to persuade his readers that the environment can help man recover his lost sense of intimacy, sensitivity, holiness, mystery, and wonder. Alan Watts, one nonscientist in this section, seeks to bring Eastern philosophy to bear on the controversy. Is the result a modern version of Emerson? The historian Theodore Roszak includes environmentalism in his romanticized "counter culture," and believes that man's salvation lies with his interconnectedness with broader mystical and natural powers in the world. Eric A. Walker proposes a more traditional alternative to an antitechnological primitivism that may be emerging among some environmentalists. He believes firmly in the continued success of engineering techniques to build a better America, while acknowledging the limits of past scientific achievements. In the final selection, Herman Kahn and Anthony J. Wiener, projecting their thoughts to the year 2000, argue for continuous technological development. They believe the immanence of a real scientific breakthrough to free mankind from its material bondage.

The ecological movement has become one of the most prophetic, eloquent, and popular movements in recent American history. It has gained advocates and disciples from the left, right, and middle of American society. It has already had two unexpected effects upon American life, although its future course and effectiveness is unclear: (1) The ecological perspective redefines human nature in terms of harmony with the natural world rather than mastery. This point of view contradicts both overt and unconscious beliefs that growth and progress in American history are good and necessary. To some environmentalists, American economic development has meant gross and self-defeating manipulation of the American land strictly for narrow human benefit. (2) With the exception of the last two selections, the authors represented in this volume do not see any effective solution to the crisis through future technological advances, nor do they exclusively seek return to some bucolic past. Rather they argue for a change in man himself, from predator and outlaw in the environment to member and citizen of a larger community, human and nonhuman, than ever before acknowledged in American life.

Conflict of Opinion

Only man among all forces in nature is wantonly destructive:

> Man pursues his victims with reckless destructiveness; and, while the sacrifice of life by the lower animals is limited by the cravings of appetite, he unsparingly persecutes, even to extirpation, thousands of organic forms which he cannot consume. . . . Purely untutored humanity, it is true, interferes comparatively little with the arrangements of nature, and the destructive agency of man becomes more and more energetic and unsparing as he advances in civilization.
> GEORGE PERKINS MARSH, 1874

Mankind can simultaneously save nature and use its resources for the development of civilization:

> There may be just as much waste in neglecting the development and use of certain natural resources as there is in their destruction. . . . The first duty of the human race is to control the earth it lives upon.
> GIFFORD PINCHOT, 1910

The modern controversy opened with a plea to recognize and correct the use of insecticides:

> The most alarming of all man's assaults upon the environment is the contamination of air, earth, rivers, and sea with dangerous and even lethal materials. This pollution is for the most part irrecoverable; the chain of evil it initiates not only in the world that must support life, but in living tissues is for the most part irreversible. In this now universal contamination of the environment, chemicals are the sinister and little-recognized partners of radiation in changing the very nature of the world—the very nature of its life.
> RACHEL CARSON, 1962

And the response was immediate, direct, and massive:

> The chemical industry presented an almost united front against what it considered the menace of Rachel Carson. . . . In November, 1962, the Manufacturing Chemists Association began mailing monthly feature stories to news media, stressing the "positive side" of chemical use. Similar material was mailed to about 100,000 individuals. . . . This was the gist of the message: "A serious threat to the continued supply of wholesome, nutritious food, and its availability at present-day low prices is manifested in the fear complex building up as a result of recent unfounded, sensational publicity with respect to agricultural chemicals."
> FRANK GRAHAM, JR., 1970

An anthropologist calls for a new understanding of the relations between man and nature:

> If we examine the living universe around us which was before man and may be after him, we find two ways in which that universe, with its inhabitants, differs from the world of man: first, it is essentially a stable universe; second, its inhabitants are intensely concentrated upon their environment. They respond to it totally, and without it, or rather when they relax from it, they merely sleep. They reflect their environment but they do not alter it. . . . It is with the coming of man that a vast hole seems to open in nature, a vast black whirlpool spinning faster and faster, consuming flesh, stones, soil, minerals, sucking down the lightning, wrenching power from the atom, until the ancient sounds of nature are drowned in the cacophony of something which is no longer nature, something instead which is loose and knocking at the world's heart.
>
> LOREN EISELEY, 1960

But two national resource planners are optimistic about the future of industrial society:

> To speak, to use tools, to pass learning on to children; to put fire, domestic animals, wind, falling water, and other energy sources to human use; to gather food, fuel, clothing, and seeds for winter; to save, invest, plan, build, and innovate in order to decrease dangers and insecurities and to increase the power to change natural things to suit one's purposes; in sum, to subdue Nature, and render her subject to human will—such have been the results, if not always the conscious goals, of eons of striving. Success would seem to be at hand; as we approach the beginning of the twenty-first century, our capacities for and commitment to economic development and technological control over our external and internal environment, as well as the concomitant systematic innovation, application, and diffusion of these capacities, seem to be increasing, and without foreseeable limit.
>
> HERMAN KAHN AND ANTHONY J. WIENER, 1968

I EARLIER AMERICAN OPINIONS ABOUT NATURE AND WILDERNESS

Ralph Waldo Emerson

THE ROMANTIC PHILOSOPHY OF NATURE

Ralph Waldo Emerson (1803–1882) was a leading exponent of Transcendentalism, an American version of English and European Romanticism. The following selection is from Nature, *published in 1836, and Emerson's first book. Like his contemporary Henry David Thoreau, he searched out the natural world for a new piety and gentler morality to replace religious orthodoxy and cold rationalism. Thoreau may have been more successful in keeping physical nature and transcendental ideals together, but Emerson claimed to have found a more profound mystical union between man and nature by identifying the essence of both with fundamental creative powers in the cosmos. Wild nature, he argued, possessed more meaning than mankind or civilization alone; it was the ultimate source of life and culture. Whoever was captured by such a vision would have the secrets of the universe as his most intimate possessions. Emerson transcendentalized nature by making it into a symbolic representation of the Oversoul, the center of universal consciousness. Nature's materiality was of secondary importance to thought. Its real significance lay in its abilities to carry man beyond ordinary rationality into inspired apprehensions and feelings. This praise of the mystical inspiration that nature can provide at the price of its own organic values was, however, what later ecologists seriously questioned. Is this a classic dualism which still valued immaterial "spirit" far more than physical nature? While Emerson's views were hardly scientific, he contributed to the persistent American belief that the natural world, once its secrets were unlocked, was a highly rewarding arena of investigation.*

To go into solitude, a man needs to retire as much from his chamber as from society. I am not solitary whilst I read and write, though nobody is with me. But if a man would be alone, let him look at the stars. The rays that come from those heavenly worlds will separate between him and what he touches. One might think the atmosphere was made transparent with this design, to give man, in the heavenly bodies, the perpetual presence of the sublime. Seen in the streets of cities, how great they are! If the stars should appear one night in a thousand years, how would men believe and adore; and preserve for many generations the remembrance of the city of God which had

From *The Complete Works of Ralph Waldo Emerson,* edited by Edward Waldo Emerson (Houghton Mifflin Co., 1903–1904).

been shown! But every night come out these envoys of beauty, and light the universe with their admonishing smile.

The stars awaken a certain reverence, because though always present, they are inaccessible; but all natural objects make a kindred impression, when the mind is open to their influence. Nature never wears a mean appearance. Neither does the wisest man extort her secret, and lose his curiosity by finding out all her perfection. Nature never became a toy to a wise spirit. The flowers, the animals, the mountains, reflected the wisdom of his best hour, as much as they had delighted the simplicity of his childhood.

When we speak of nature in this manner, we have a distinct but most poetical sense in the mind. We mean the integrity of impression made by manifold natural objects. It is this which distinguishes the stick of timber of the wood-cutter from the tree of the poet. The charming landscape which I saw this morning is indubitably made up of some twenty or thirty farms. Miller owns this field, Locke that, and Manning the woodland beyond. But none of them owns the landscape. There is a property in the horizon which no man has but he whose eye can integrate all the parts, that is, the poet. This is the best part of these men's farms, yet to this their warranty-deeds give no title.

To speak truly, few adult persons can see nature. Most persons do not see the sun. At least they have a very superficial seeing. The sun illuminates only the eye of the man, but shines into the eye and the heart of the child. The lover of nature is he whose inward and outward senses are still truly adjusted to each other; who has retained the spirit of infancy even into the era of manhood. His intercourse with heaven and earth becomes part of his daily food. In the presence of nature a wild delight runs through the man, in spite of real sorrows. Nature says—he is my creature, and maugre all his impertinent griefs, he shall be glad with me. Not the sun or the summer alone, but every hour and season yields its tribute of delight; for every hour and change corresponds to and authorizes a different state of the mind, from breathless noon to grimmest midnight. Nature is a setting that fits equally well a comic or a mourning piece. In good health, the air is a cordial of incredible virtue. Crossing a bare common, in snow puddles, at twilight, under a clouded sky, without having in my thoughts any occurrence of special good fortune, I have enjoyed a perfect exhilaration. I am glad to the brink of fear. In the woods, too, a man casts off his years, as the snake his slough, and at what period soever of life is always a child. In the woods is perpetual youth. Within these plantations of God, a decorum and sanctity reign, a perennial festival is dressed, and the guest sees not how he should tire of them in a thousand years. In the woods, we return to reason and faith. There I feel that nothing can befall me in life—no disgrace, no calamity (leaving me my eyes), which nature cannot repair. Standing on the bare ground—my head bathed by the blithe air and uplifted into infinite space—all mean egotism vanishes. I become a transparent eyeball; I am nothing; I see all; the currents of the Universal Being circulate through me; I am part or parcel of God. The name of the nearest friend sounds then foreign and accidental: to be brothers, to be acquaintances, master or servant, is then a trifle and a disturbance. I am the lover of uncontained and immortal beauty. In the wilderness, I find something more dear and connate than in streets or villages. In the tranquil landscape, and especially in the distant line of the horizon, man beholds somewhat as beautiful as his own nature.

The greatest delight which the fields and woods minister is the suggestion of an occult relation between man and the vegetable. I am not alone and unacknowledged. They nod to me, and I to them. The waving of the boughs in the storm is new to me and old. It takes me by surprise, and yet is not unknown. Its effect is like that of a higher thought or a better emotion coming over me, when I deemed I was thinking justly or doing right.

Yet it is certain that the power to produce this delight does not reside in nature, but in man, or in a harmony of both. It is necessary to use these pleasures with great temperance. For nature is not always tricked in holiday attire, but the same scene which yesterday breathed perfume and glittered as for the frolic of the nymphs is overspread with melancholy to-day. Nature always wears the colors of the spirit. To a man laboring under calamity, the heat of his own fire hath sadness in it. Then there is a kind of contempt of the landscape felt by him who has just lost by death a dear friend. The sky is less grand as it shuts down over less worth in the population.

*   *   *

. . . The misery of man appears like childish petulance, when we explore the steady and prodigal provision that has been made for his support and delight on this green ball which floats him through the heavens. What angels invented these splendid ornaments, these rich conveniences, this ocean of air above, this ocean of water beneath, this firmament of earth between? this zodiac of lights, this tent of dropping clouds, this striped coat of climates, this fourfold year? Beasts, fire, water, stones, and corn serve him. The field is at once his floor, his work-yard, his play-ground, his garden, and his bed.

> *More servants wait on man*
> *Than he'll take notice of.**

Nature, in its ministry to man, is not only the material, but is also the process and the result. All the parts incessantly work into each other's hands for the profit of man. The wind sows the seed; the sun evaporates the sea; the wind blows the vapor to the field; the ice, on the other side of the planet, condenses rain on this; the rain feeds the plant; the plant feeds the animal; and thus the endless circulations of the divine charity nourish man.

*   *   *

A nobler want of man is served by nature, namely, the love of Beauty.

The ancient Greeks called the world κόσμος, beauty. Such is the constitution of all things, or such the plastic power of the human eye, that the primary forms, as the sky, the mountain, the tree, the animal, give us a delight *in and for themselves;* a pleasure arising from outline, color, motion, and grouping. This seems partly owing to the eye itself. The eye is the best of artists. By the mutual action of its structure and of the laws of light, perspective is produced, which integrates every mass of objects, of what character soever, into a well colored and shaded globe, so that where the particular objects are mean and unaffecting, ·the landscape which they compose is round and symmetrical. And as the eye is the best composer,

---

* George Herbert's poem "Man."

so light is the first of painters. There is no object so foul that intense light will not make beautiful. And the stimulus it affords to the sense, and a sort of infinitude which it hath, like space and time, make all matter gay. Even the corpse has its own beauty. But besides this general grace diffused over nature, almost all the individual forms are agreeable to the eye, as is proved by our endless imitations of some of them, as the acorn, the grape, the pine-cone, the wheat-ear, the egg, the wings and forms of most birds, the lion's claw, the serpent, the butterfly, sea-shells, flames, clouds, buds, leaves, and the forms of many trees, as the palm.

. . . The simple perception of natural forms is a delight. The influence of the forms and actions in nature is so needful to man, that, in its lowest functions, it seems to lie on the confines of commodity and beauty. To the body and mind which have been cramped by noxious work or company, nature is medicinal and restores their tone. The tradesman, the attorney comes out of the din and craft of the street and sees the sky and the woods, and is a man again. In their eternal calm, he finds himself. The health of the eye seems to demand a horizon. We are never tired, so long as we can see far enough.

*   *   *

. . . We know more from nature than we can at will communicate. Its light flows into the mind evermore, and we forget its presence. The poet, the orator, bred in the woods, whose senses have been nourished by their fair and appeasing changes, year after year, without design and without heed—shall not lose their lesson altogether, in the roar of cities or the broil of politics. Long hereafter amidst agitation and terror in national councils—in the hour of revolution—these solemn images shall reappear in their morning lustre, as fit symbols and words of the thoughts which the passing events shall awaken. At the call of a noble sentiment, again the woods wave, the pines murmur, the river rolls and shines, and the cattle low upon the mountains, as he saw and heard them in his infancy. And with these forms, the spells of persuasion, the keys of power are put into his hands.

. . . We are thus assisted by natural objects in the expression of particular meanings. But how great a language to convey such pepper-corn informations! Did it need such noble races of creatures, this profusion of forms, this host of orbs in heaven, to furnish man with the dictionary and grammar of his municipal speech? Whilst we use this grand cipher to expedite the affairs of our pot and kettle, we feel that we have not yet put it to its use, neither are able. We are like travellers using the cinders of a volcano to roast their eggs. Whilst we see that it always stands ready to clothe what we would say, we cannot avoid the question whether the characters are not significant of themselves. Have mountains, and waves, and skies, no significance but what we consciously give them when we employ them as emblems of our thoughts? The world is emblematic. Parts of speech are metaphors, because the whole of nature is a metaphor of the human mind.

*   *   *

. . . What good heed Nature forms in us! She pardons no mistakes. Her yea is yea, and her nay, nay.

The first steps in Agriculture, Astronomy, Zoölogy (those first steps which the farmer, the hunter, and the sailor take), teach that Nature's dice are always loaded; that in her heaps and rubbish are concealed sure and useful results.

How calmly and genially the mind apprehends one after another the laws of physics! What noble emotions dilate the mortal as he enters into the councils of the creation, and feels by knowledge the privilege to *Be!* His insight refines him. The beauty of nature shines in his own breast. Man is greater that he can see this, and the universe less, because Time and Space relations vanish as laws are known.

Here again we are impressed and even daunted by the immense Universe to be explored. "What we know is a point to what we do not know." Open any recent journal of science, and weigh the problems suggested concerning Light, Heat, Electricity, Magnetism, Physiology, Geology, and judge whether the interest of natural science is likely to be soon exhausted.

. . . Nature is thoroughly mediate. It is made to serve. It receives the dominion of man as meekly as the ass on which the Saviour rode. It offers all its kingdoms to man as the raw material which he may mould into what is useful. Man is never weary of working it up. He forges the subtile and delicate air into wise and melodious words, and gives them wing as angels of persuasion and command. One after another his victorious thought comes up with and reduces all things, until the world becomes at last only a realized will—the double of the man.

*   *   *

. . . The moral law lies at the centre of nature and radiates to the circumference. It is the pith and marrow of every substance, every relation, and every process. All things with which we deal, preach to us. What is a farm but a mute gospel? The chaff and the wheat, weeds and plants, blight, rain, insects, sun—it is a sacred emblem from the first furrow of spring to the last stack which the snow of winter overtakes in the fields. But the sailor, the shepherd, the miner, the merchant, in their several resorts, have each an experience precisely parallel, and leading to the same conclusion: because all organizations are radically alike. Nor can it be doubted that this moral sentiment which thus scents the air, grows in the grain, and impregnates the waters of the world, is caught by man and sinks into his soul. The moral influence of nature upon every individual is that amount of truth which it illustrates to him. Who can estimate this? Who can guess how much firmness the sea-beaten rock has taught the fisherman? how much tranquillity has been reflected to man from the azure sky, over whose unspotted deeps the winds forevermore drive flocks of stormy clouds, and leave no wrinkle or stain? how much industry and providence and affection we have caught from the pantomime of brutes? What a searching preacher of self-command is the varying phenomenon of Health!

Herein is especially apprehended the unity of Nature—the unity in variety—which meets us everywhere. All the endless variety of things make an identical impression. Xenophanes complained in his old age, that, look where he would, all things hastened back to Unity. He was weary of seeing the same entity in the tedious variety of forms. The fable of Proteus has a cordial truth. A leaf, a drop, a crystal, a moment of time, is related to the whole, and partakes of the perfection of the whole. Each particle is a microcosm, and faithfully renders the likeness of the world.

*   *   *

I shall therefore conclude this essay with some traditions of man and nature, which a certain poet sang to me; and which, as they have always been in the world, and perhaps reappear to every bard, may be both history and prophecy.

>   *The foundations of man are not in matter, but in spirit. But the element of spirit is eternity. To it, therefore, the longest series of events, the oldest chronologies are young and recent. In the cycle of the universal man, from whom the known individuals proceed, centuries are points, and all history is but the epoch of one degradation.*
>
>   *We distrust and deny inwardly our sympathy with nature. We own and disown our relation to it, by turns. We are like Nebuchadnezzar, dethroned, bereft of reason, and eating like an ox. But who can set limits to the remedial force of spirit?*

*   *   *

At present, man applies to nature but half his force. He works on the world with his understanding alone. He lives in it and masters it by a penny-wisdom; and he that works most in it is but a half-man, and whilst his arms are strong and his digestion good, his mind is imbruted, and he is a selfish savage. His relation to nature, his power over it, is through the understanding, as by manure; the economic use of fire, wind, water, and the mariner's needle; steam, coal, chemical agriculture; the repairs of the human body by the dentist and the surgeon. This is such a resumption of power as if a banished king should buy his territories inch by inch, instead of vaulting at once into his throne.

*   *   *

So shall we come to look at the world with new eyes. It shall answer the endless inquiry of the intellect—What is truth? and of the affections—What is good? by yielding itself passive to the educated Will. Then shall come to pass what my poet said:

>   *Nature is not fixed but fluid. Spirit alters, moulds, makes it. The immobility or bruteness of nature is the absence of spirit; to pure spirit it*

is fluid, it is volatile, it is obedient. Every spirit builds itself a house, and beyond its house a world, and beyond its world a heaven. Know then that the world exists for you. For you is the phenomenon perfect. What we are, that only can we see. All that Adam had, all that Cæsar could, you have and can do. Adam called his house, heaven and earth; Cæsar called his house, Rome; you perhaps call yours, a cobbler's trade; a hundred acres of ploughed land; or a scholar's garret. Yet line for line and point for point your dominion is as great as theirs, though without fine names. Build therefore your own world. As fast as you conform your life to the pure idea in your mind, that will unfold its great proportions. A correspondent revolution in things will attend the influx of the spirit. So fast will disagreeable appearances, swine, spiders, snakes, pests, mad-houses, prisons, enemies, vanish; they are temporary and shall be no more seen. The sordor and filths of nature, the sun shall dry up and the wind exhale. As when the summer comes from the south the snow-banks melt and the face of the earth becomes green before it, so shall the advancing spirit create its ornaments along its path, and carry with it the beauty it visits and the song which enchants it; it shall draw beautiful faces, warm hearts, wise discourse, and heroic acts, around its way, until evil is no more seen. The kingdom of man over nature, which cometh not with observation—a dominion such as now is beyond his dream of God—he shall enter without more wonder than the blind man feels who is gradually restored to perfect sight.

George Perkins Marsh

MAN AS PREDATOR ON THE EARTH

George Perkins Marsh (1801–1882) was a contemporary of Emerson, but he wrote much later about the environment, and in a surprisingly modern vein. Also a Massachusetts man, Perkins was a successful businessman, lawyer, scholar, linguist, congressman, and diplomat to Turkey and Italy. This varied and international background led him to place American civilization in a cosmopolitan context and encouraged him to consider the different possibilities for human dealings with the natural world. Man and Nature; or Phys-

From *The Earth as Modified by Human Action. A New Edition of Man and Nature* by George Perkins Marsh (Scribner, Armstrong and Co., 1874). Notes to the original have been omitted.

ical Geography as Modified by Human Action, first published in 1864, is now recognized as a pioneering work. More than any other nineteenth-century writer, Marsh anticipated the crusading spirit of today's ecologists to overcome man's threat to the welfare and survival of the environment. Marsh believed that the destruction of the natural world inevitably doomed civilization. Empires fell when their natural resources were wasted, and he made a special case against the wasteful use of forests and woodlands in Europe and America. A strong influence on the views of John Muir, Marsh was one of the first Americans who believed that nature was not an infinite resource to be used indiscriminately. The following selection is taken from Marsh's revised edition of 1874.

Reaction of Man on Nature

The revolutions of the seasons, with their alternations of temperature and of length of day and night, the climates of different zones, and the general conditions and movements of the atmosphere and the seas, depend upon causes for the most part cosmical, and, of course, wholly beyond our control. The elevation, configuration, and composition of the great masses of terrestrial surface, and the relative extent and distribution of land and water, are determined by geological influences equally remote from our jurisdiction. It would hence seem that the physical adaptation of different portions of the earth to the use and enjoyment of man is a matter so strictly belonging to mightier than human powers, that we can only accept geographical nature as we find her, and be content with such soils and such skies as she spontaneously offers.

But it is certain that man has reacted upon organized and inorganic nature, and thereby modified, if not determined, the material structure of his earthly home. The measure of that reaction manifestly constitutes a very important element in the appreciation of the relations between mind and matter, as well as in the discussion of many purely physical problems. But though the subject has been incidentally touched upon by many geographers, and treated with much fulness of detail in regard to certain limited fields of human effort and to certain specific effects of human action, it has not, as a whole, so far as I know, been made matter of special observation, or of historical research, by any scientific inquirer. Indeed, until the influence of geographical conditions upon human life was recognized as a distinct branch of philosophical investigation, there was no motive for the pursuit of such speculations; and it was desirable to inquire how far we have, or can, become the architects of our own abiding place, only when it was known how the mode of our physical, moral, and intellectual being is affected by the character of the home which Providence has appointed, and we have fashioned, for our material habitation.

It is still too early to attempt scientific method in discussing this problem, nor is our present store of the necessary facts by any means complete enough to warrant me in promising any approach to fulness of statement respecting them. Systematic observation in relation to this subject has hardly yet begun, and the scattered data which have chanced to be recorded have never been collected. It has now no place in the general scheme of physical science, and is matter of suggestion and speculation only, not of established and positive conclusion. At present, then, all that I can hope is to excite an interest in a topic of much economical importance, by pointing out the directions and illustrating the modes in which human action has been, or may be, most injurious or most beneficial in its influence upon the physical conditions of the earth we inhabit.

We cannot always distinguish between the results of man's action and the effects of purely geological or cosmical causes. The destruction of the forests, the drainage of lakes and marshes, and the operations of rural husbandry and industrial art have unquestionably tended to produce great changes in the hygrometric, thermometric, electric, and chemical condition of the atmosphere, though we are not yet able to measure the force of the different elements of disturbance, or to say how far they have been neutralised by each other, or by still obscurer influences; and it is equally certain that the myriad forms of animal and vegetable life, which covered the earth when man first entered upon the theatre of a nature whose harmonies he was destined to derange, have been, through his interference, greatly changed in numerical proportion, sometimes much modified in form and product, and sometimes entirely extirpated.

The physical revolutions thus wrought by man have not indeed all been destructive to human interests, and the heaviest blows he has inflicted upon nature have not been wholly without their compensations. Soils to which no nutritious vegetable was indigenous, coun-

tries which once brought forth but the fewest products suited for the
sustenance and comfort of man—while the severity of their climates
created and stimulated the greatest number and the most imperious
urgency of physical wants—surfaces the most rugged and intrac-
table, and least blessed with natural facilities of communication,
have been brought in modern times to yield and distribute all that
supplies the material necessities, all that contributes to the sensuous enjoyments and conveniences of civilized life. The Scythia,
the Thule, the Britain, the Germany, and the Gaul which the Roman
writers describe in such forbidding terms, have been brought almost
to rival the native luxuriance and easily won plenty of Southern
Italy; and, while the fountains of oil and wine that refreshed old
Greece and Syria and Northern Africa have almost ceased to flow,
and the soils of those fair lands are turned to thirsty and inhospi-
table deserts, the hyperborean regions of Europe have learned to
conquer, or rather compensate, the rigors of climate, and have
attained to a material wealth and variety of product that, with all
their natural advantages, the granaries of the ancient world can
hardly be said to have enjoyed.

*   *   *

Stability of Nature

Nature, left undisturbed, so fashions her territory as to give it
almost unchanging permanence of form, outline, and proportion,
except when shattered by geologic convulsions; and in these comparatively rare cases of derangement, she sets herself at once to
repair the superficial damage, and to restore, as nearly as practi-
cable, the former aspect of her dominion. In new countries, the
natural inclination of the ground, the self-formed slopes and levels,
are generally such as best secure the stability of the soil. They have
been graded and lowered or elevated by frost and chemical forces
and gravitation and the flow of water and vegetable deposit and the
action of the winds, until, by a general compensation of conflicting
forces, a condition of equilibrium has been reached which, without
the action of man, would remain, with little fluctuation, for countless
ages.

We need not go far back to reach a period when, in all that

portion of the North American continent which has been occupied by British colonization, the geographical elements very nearly balanced and compensated each other. At the commencement of the seventeenth century, the soil, with insignificant exceptions, was covered with forests; and whenever the Indian, in consequence of war or the exhaustion of the beasts of the chase, abandoned the narrow fields he had planted and the woods he had burned over, they speedily returned, by a succession of herbaceous, arborescent, and arboreal growths, to their original state. Even a single generation sufficed to restore them almost to their primitive luxuriance of forest vegetation. The unbroken forests had attained to their maximum density and strength of growth, and, as the older trees decayed and fell, they were succeeded by new shoots or seedlings, so that from century to century no perceptible change seems to have occurred in the wood, except the slow, spontaneous succession of crops. This succession involved no interruption of growth, and but little break in the "boundless contiguity of shade"; for, in the husbandry of nature, there are no fallows. Trees fall singly, not by square roods, and the tall pine is hardly prostrate, before the light and heat, admitted to the ground by the removal of the dense crown of foliage which had shut them out, stimulate the germination of the seeds of broad-leaved trees that had lain, waiting this kindly influence, perhaps for centuries.

* * *

Destructiveness of Man

Man has too long forgotten that the earth was given to him for usufruct alone, not for consumption, still less for profligate waste. Nature has provided against the absolute destruction of any of her elementary matter, the raw material of her works; the thunderbolt and the tornado, the most convulsive throes of even the volcano and the earthquake, being only phenomena of decomposition and recomposition. But she has left it within the power of man irreparably to derange the combinations of inorganic matter and of organic life, which through the night of æons she had been proportioning and balancing, to prepare the earth for his habitation, when in the fulness of time his Creator should call him forth to enter into its possession.

Apart from the hostile influence of man, the organic and the in-organic world are, as I have remarked, bound together by such mutual relations and adaptations as secure, if not the absolute per-manence and equilibrium of both, a long continuance of the estab-lished conditions of each at any given time and place, or at least, a very slow and gradual succession of changes in those conditions. But man is everywhere a disturbing agent. Wherever he plants his foot, the harmonies of nature are turned to discords. The proportions and accommodations which insured the stability of existing arrange-ments are overthrown. Indigenous vegetable and animal species are extirpated, and supplanted by others of foreign origin, spontaneous production is forbidden or restricted, and the face of the earth is either laid bare or covered with a new and reluctant growth of vege-table forms, and with alien tribes of animal life. These intentional changes and substitutions constitute, indeed, great revolutions; but vast as is their magnitude and importance, they are, as we shall see, insignificant in comparison with the contingent and unsought results which have flowed from them.

The fact that, of all organic beings, man alone is to be regarded as essentially a destructive power, and that he wields energies to resist which Nature—that nature whom all material life and all inor-ganic substance obey—is wholly impotent, tends to prove that, though living in physical nature, he is not of her, that he is of more exalted parentage, and belongs to a higher order of existences, than those which are born of her womb and live in blind submission to her dictates.

There are, indeed, brute destroyers, beasts and birds and insects of prey—all animal life feeds upon, and, of course, destroys other life,—but this destruction is balanced by compensations. It is, in fact, the very means by which the existence of one tribe of animals or of vegetables is secured against being smothered by the en-croachments of another; and the reproductive powers of species, which serve as the food of others, are always proportioned to the demand they are destined to supply. Man pursues his victims with reckless destructiveness; and, while the sacrifice of life by the lower animals is limited by the cravings of appetite, he unsparingly perse-cutes, even to extirpation, thousands of organic forms which he cannot consume.

The earth was not, in its natural condition, completely adapted to the use of man, but only to the sustenance of wild animals and wild vegetation. These live, multiply their kind in just proportion, and attain their perfect measure of strength and beauty, without producing or requiring any important change in the natural arrangements of surface, or in each other's spontaneous tendencies, except such mutual repression of excessive increase as may prevent the extirpation of one species by the encroachments of another. In short, without man, lower animal and spontaneous vegetable life would have been practically constant in type, distribution, and proportion, and the physical geography of the earth would have remained undisturbed for indefinite periods, and been subject to revolution only from slow development, from possible, unknown cosmical causes, or from geological action.

But man, the domestic animals that serve him, the field and garden plants the products of which supply him with food and clothing, cannot subsist and rise to the full development of their higher properties, unless brute and unconscious nature be effectually combated, and, in a great degree, vanquished by human art. Hence, a certain measure of transformation of terrestrial surface, of suppression of natural, and stimulation of artificially modified productivity becomes necessary. This measure man has unfortunately exceeded. He has felled the forests whose network of fibrous roots bound the mould to the rocky skeleton of the earth; but had he allowed here and there a belt of woodland to reproduce itself by spontaneous propagation, most of the mischiefs which his reckless destruction of the natural protection of the soil has occasioned would have been averted. He has broken up the mountain reservoirs, the percolation of whose waters through unseen channels supplied the fountains that refreshed his cattle and fertilized his fields; but he has neglected to maintain the cisterns and the canals of irrigation which a wise antiquity had constructed to neutralize the consequences of its own imprudence. While he has torn the thin glebe which confined the light earth of extensive plains, and has destroyed the fringe of semi-aquatic plants which skirted the coast and checked the drifting of the sea sand, he has failed to prevent the spreading of the dunes by clothing them with artificially propagated vegetation. He has ruthlessly warred on all the tribes of animated nature whose spoil he could convert to his own uses, and he has not protected the birds which prey on the insects most destructive to his own harvests.

Purely untutored humanity, it is true, interferes comparatively little with the arrangements of nature, and the destructive agency of man becomes more and more energetic and unsparing as he advances in civilization, until the impoverishment, with which his exhaustion of the natural resources of the soil is threatening him, at last awakens him to the necessity of preserving what is left, if not of restoring what has been wantonly wasted. The wandering savage grows no cultivated vegetable, fells no forest, and extirpates no useful plant, no noxious weed. If his skill in the chase enables him to entrap numbers of the animals on which he feeds, he compensates this loss by destroying also the lion, the tiger, the wolf, the otter, the seal, and the eagle, thus indirectly protecting the feebler quadrupeds and fish and fowls, which would otherwise become the booty of beasts and birds of prey. But with stationary life, or at latest with the pastoral state, man at once commences an almost indiscriminate warfare upon all the forms of animal and vegetable existence around him, and as he advances in civilization, he gradually eradicates or transforms every spontaneous product of the soil he occupies.

Human and Brute Action Compared

It is maintained by authorities as high as any known to modern science, that the action of man upon nature, though greater in *degree,* does not differ in *kind* from that of wild animals. It is perhaps impossible to establish a radical distinction *in genere* between the two classes of effects, but there is an essential difference between the motive of action which calls out the energies of civilized man and the mere appetite which controls the life of the beast. The action of man, indeed, is frequently followed by unforeseen and undesired results, yet it is nevertheless guided by a self-conscious will aiming as often at secondary and remote as at immediate objects. The wild animal, on the other hand, acts instinctively, and, so far as we are able to perceive, always with a view to single and direct purposes. The backwoodsman and the beaver alike fell trees;

the man that he may convert the forest into an olive grove that will mature its fruit only for a succeeding generation, the beaver that he may feed upon the bark of the trees or use them in the construction of his habitation. The action of brutes upon the material world is slow and gradual, and usually limited, in any given case, to a narrow extent of territory. Nature is allowed time and opportunity to set her restorative powers at work, and the destructive animal has hardly retired from the field of his ravages before nature has repaired the damages occasioned by his operations. In fact, he is expelled from the scene by the very efforts which she makes for the restoration of her dominion. Man, on the contrary, extends his action over vast spaces, his revolutions are swift and radical, and his devastations are, for an almost incalculable time after he has withdrawn the arm that gave the blow, irreparable.

The form of geographical surface, and very probably the climate of a given country, depend much on the character of the vegetable life belonging to it. Man has, by domestication, greatly changed the habits and properties of the plants he rears; he has, by voluntary selection, immensely modified the forms and qualities of the animated creatures that serve him; and he has, at the same time, completely rooted out many forms of animal if not of vegetable being. What is there, in the influence of brute life, that corresponds to this? We have no reason to believe that, in that portion of the American continent which, though peopled by many tribes of quadruped and fowl, remained uninhabited by man or only thinly occupied by purely savage tribes, any sensible geographical change had occurred within twenty centuries before the epoch of discovery and colonization, while, during the same period, man had changed millions of square miles, in the fairest and most fertile regions of the Old World, into the barrenest deserts.

The ravages committed by man subvert the relations and destroy the balance which nature had established between her organized and her inorganic creations, and she avenges herself upon the intruder, by letting loose upon her defaced provinces destructive energies hitherto kept in check by organic forces destined to be his best auxiliaries, but which he has unwisely dispersed and driven from the field of action. When the forest is gone, the great reservoir of moisture stored up in its vegetable mould is evaporated, and returns only in deluges of rain to wash away the parched dust into which that mould has been converted. The well-wooded and humid hills are turned to ridges of dry rock, which encumbers the low grounds and chokes the watercourses with its debris, and—except in countries favored with an equable distribution of rain through the seasons, and a moderate and regular inclination of surface— the whole earth, unless rescued by human art from the physical degradation to which it tends, becomes an assemblage of bald mountains, of barren, turfless hills, and of swampy and malarious plains. There are parts of Asia Minor, of Northern Africa, of Greece, and even of Alpine Europe, where the operation of causes set in action by man has brought the face of the earth to a desolation almost as complete as that of the moon; and though, within that brief space of time which we call "the historical period," they are known to have been covered with luxuriant woods, verdant pastures, and fertile meadows, they are now too far deteriorated to be reclaimable by man, nor can they become again fitted for human use, except through great geological changes, or other mysterious influences or agencies of which we have no present knowledge, and over which we have no prospective control. The earth is fast becoming an unfit home for its noblest inhabitant, and another era of equal human crime and human improvidence, and of like duration with that through which traces of that crime and that improvidence extend, would reduce it to such a condition of impoverished productiveness, of shattered surface, of climatic excess, as to threaten the depravation, barbarism, and perhaps even extinction of the species.

Physical Improvement

True, there is a partial reverse to this picture. On narrow theatres, new forests have been planted; inundations of flowing streams restrained by heavy walls of masonry and other constructions; torrents compelled to aid, by depositing the slime with which they are charged, in filling up lowlands, and raising the level of morasses which their own overflows had created; ground submerged by the encroachments of the ocean, or exposed to be covered by its tides, has been rescued from its dominion by diking; swamps and even lakes have been drained, and their beds brought within the domain of agricultural industry; drifting coast dunes have been checked and made productive by plantation; seas and inland waters have been repeopled with fish, and even the sands of the Sahara have been fertilized by artesian fountains. These achievements are more glorious than the proudest triumphs of war, but, thus far, they give but faint hope that we shall yet make full atonement for our spend-thrift waste of the bounties of nature.

Limits of Human Power

It is, on the one hand, rash and unphilosophical to attempt to set limits to the ultimate power of man over inorganic nature, and it is unprofitable, on the other, to speculate on what may be accomplished by the discovery of now unknown and unimagined natural forces, or even by the invention of new arts and new processes. But since we have seen aerostation, the motive power of elastic vapors, the wonders of modern telegraphy, the destructive explosiveness of gunpowder, of nitro-glycerine, and even of a substance so harmless, unresisting, and inert as cotton, there is little in the way of mechanical achievement which seems hopelessly impossible, and it is hard to restrain the imagination from wandering forward a couple of generations to an epoch when our descendants shall have advanced as far beyond us in physical conquest, as we have marched beyond the trophies erected by our grandfathers. There are, nevertheless, in actual practice, limits to the efficiency of the forces which we are now able to bring into the field, and we must admit that, for the present, the agencies known to man and controlled by him are inadequate to the reducing of great Alpine precipices to such slopes as would enable them to support a vegetable clothing, or to the covering of large extents of denuded rock with earth, and planting upon them a forest growth. Yet among the mysteries which science is hereafter to reveal, there may be still undiscovered methods of accomplishing even grander wonders than these. Mechanical philosophers have suggested the possibility of accumulating and treasuring up for human use some of the greater natural forces, which the action of the elements puts forth with such astonishing energy.

Could we gather, and bind, and make subservient to our control, the power which a West Indian hurricane exerts through a small area in one continuous blast, or the momentum expended by the waves, in a tempestuous winter, upon the breakwater at Cherbourg, or the lifting power of the tide, for a month, at the head of the Bay of Fundy, or the pressure of a square mile of sea water at the depth of five thousand fathoms, or a moment of the might of an earth-quake or a volcano, our age—which moves no mountains and casts them into the sea by faith alone—might hope to scarp the rugged walls of the Alps and Pyrenees and Mount Taurus, robe them once more in a vegetation as rich as that of their pristine woods, and turn their wasting torrents into refreshing streams.

Could this old world, which man has overthrown, be rebuilded, could human cunning rescue its wasted hillsides and its deserted plains from solitude or mere nomade occupation, from barrenness, from nakedness, and from insalubrity, and restore the ancient fertil-ity and healthfulness of the Etruscan sea coast, the Campagna and the Pontine marshes, of Calabria, of Sicily, of the Peloponnesus and insular and continental Greece, of Asia Minor, of the slopes of Lebanon and Hermon, of Palestine, of the Syrian desert, of Meso-potamia and the delta of the Euphrates, of the Cyrenaica, of Africa proper, Numidia, and Mauritania, the thronging millions of Europe might still find room on the Eastern continent, and the main current of emigration be turned towards the rising instead of the setting sun.

But changes like these must await not only great political and moral revolutions in the governments and peoples by whom those regions are now possessed, but, especially, a command of pecuni-ary and of mechanical means not at present enjoyed by those na-tions, and a more advanced and generally diffused knowledge of the processes by which the amelioration of soil and climate is possi-ble than now anywhere exists. Until such circumstances shall con-spire to favor the work of geographical regeneration, the countries I have mentioned, with here and there a local exception, will con-tinue to sink into yet deeper desolation, and in the meantime the American continent, Southern Africa, Australia, New Zealand, and the smaller oceanic islands, will be almost the only theatres where man is engaged, on a great scale, in transforming the face of nature.

Importance of Physical Conservation and Restoration

Comparatively short as is the period through which the colonization of foreign lands by European emigrants extends, great and, it
is to be feared, sometimes irreparable injury has already been done
in the various processes by which man seeks to subjugate the virgin
earth; and many provinces, first trodden by the *homo sapiens Eu-
ropæ* within the last two centuries, begin to show signs of that melancholy dilapidation which is now driving so many of the peasantry
of Europe from their native hearths. It is evidently a matter of great
moment, not only to the population of the states where these symptoms are manifesting themselves, but to the general interests of
humanity, that this decay should be arrested, and that the future
operations of rural husbandry and of forest industry, in districts yet
remaining substantially in their native condition, should be so conducted as to prevent the widespread mischiefs which have been
elsewhere produced by thoughtless or wanton destruction of the
natural safeguards of the soil. This can be done only by the diffusion of knowledge on this subject among the classes that, in earlier
days, subdued and tilled ground in which they had no vested rights,
but who, in our time, own their woods, their pastures, and their
ploughlands as a perpetual possession for them and theirs, and
have, therefore, a strong interest in the protection of their domain
against deterioration.

Physical Restoration

Many circumstances conspire to invest with great present interest
the questions: how far man can permanently modify and ameliorate
those physical conditions of terrestrial surface and climate on which
his material welfare depends; how far he can compensate, arrest,
or retard the deterioration which many of his agricultural and industrial processes tend to produce; and how far he can restore fertility and salubrity to soils which his follies or his crimes have made
barren or pestilential. Among these circumstances, the most prominent, perhaps, is the necessity of providing new homes for a European population which is increasing more rapidly than its means
of subsistence, new physical comforts for classes of the people that

have now become too much enlightened and have imbibed too much culture to submit to a longer deprivation of a share in the material enjoyments which the privileged ranks have hitherto monopolized.

To supply new hives for the emigrant swarms, there are, first, the vast unoccupied prairies and forests of America, of Australia, and of many other great oceanic islands, the sparsely inhabited and still unexhausted soils of Southern and even Central Africa, and, finally, the impoverished and half-depopulated shores of the Mediterranean, and the interior of Asia Minor and the farther East. To furnish to those who shall remain after emigration shall have conveniently reduced the too dense population of many European states, those means of sensuous and of intellectual well-being which are styled "artificial wants" when demanded by the humble and the poor, but are admitted to be "necessaries" when claimed by the noble and the rich, the soil must be stimulated to its highest powers of production, and man's utmost ingenuity and energy must be tasked to renovate a nature drained, by his improvidence, of fountains which a wise economy would have made plenteous and perennial sources of beauty, health, and wealth.

In those yet virgin lands which the progress of modern discovery in both hemispheres has brought and is still bringing to the knowledge and control of civilized man, not much improvement of great physical conditions is to be looked for. The proportion of forest is indeed to be considerably reduced, superfluous waters to be drawn off, and routes of internal communication to be constructed; but the primitive geographical and climatic features of these countries ought to be, as far as possible, retained.

In reclaiming and reoccupying lands laid waste by human improvidence or malice, and abandoned by man, or occupied only by a nomade or thinly scattered population, the task of the pioneer settler is of a very different character. He is to become a co-worker with nature in the reconstruction of the damaged fabric which the negligence or the wantonness of former lodgers has rendered untenantable. He must aid her in reclothing the mountain slopes with forests and vegetable mould, thereby restoring the fountains which she provided to water them; in checking the devastating fury of torrents, and bringing back the surface drainage to its primitive narrow channels; and in drying deadly morasses by opening the natural sluices which have been choked up, and cutting new canals for drawing off their stagnant waters. He must thus, on the one hand, create new reservoirs, and, on the other, remove mischievous accumulations of moisture, thereby equalizing and regulating the sources of atmospheric humidity and of flowing water, both which are so essential to all vegetable growth, and, of course, to human and lower animal life.

I have remarked that the effects of human action on the forms of the earth's surface could not always be distinguished from those resulting from geological causes, and there is also much uncertainty in respect to the precise influence of the clearing and cultivating of the ground, and of other rural operations, upon climate. It is disputed whether either the mean or the extremes of temperature, the periods of the seasons, or the amount or distribution of precipitation and of evaporation, in any country whose annals are known, have undergone any change during the historical period. It is, indeed, as has been already observed, impossible to doubt that many of the operations of the pioneer settler *tend* to produce great modifications in atmospheric humidity, temperature, and electricity; but we are at present unable to determine how far one set of effects is neutralized by another, or compensated by unknown agencies. This question scientific research is inadequate to solve, for want of the necessary data; but well conducted observation, in regions now first brought under the occupation of man, combined with such historical evidence as still exists, may be expected at no distant period to throw much light on this subject.

Australia and New Zealand are, perhaps, the countries from which we have a right to expect the fullest elucidation of these difficult and disputable problems. Their colonization did not commence until the physical sciences had become matter of almost universal attention, and is, indeed, so recent that the memory of living men embraces the principal epochs of their history; the peculiarities of their fauna, their flora, and their geology are such as to have excited for them the liveliest interest of the votaries of natural science; their mines have given their people the necessary wealth for procuring the means of instrumental observation, and the leisure required for the pursuit of scientific research; and large tracts of virgin forest and natural meadow are rapidly passing under the control of civilized man. Here, then, exist greater facilities and stronger motives for the careful study of the topics in question than have ever been found combined in any other theatre of European colonization.

In North America, the change from the natural to the artificial condition of terrestrial surface began about the period when the most important instruments of meteorological observation were invented. The first settlers in the territory now constituting the United States and the British American provinces had other things to do than to tabulate barometrical and thermometrical readings, but there remain some interesting physical records from the early days of the colonies, and there is still an immense extent of North American soil where the industry and the folly of man have as yet produced little appreciable change. Here, too, with the present increased facilities for scientific observation, the future effects, direct and contingent, of man's labors, can be measured, and such precautions taken in those rural processes which we call improvements, as to mitigate evils, perhaps, in some degree, inseparable from every attempt to control the action of natural laws.

In order to arrive at safe conclusions, we must first obtain a more exact knowledge of the topography, and of the present superficial and climatic condition of countries where the natural surface is as yet more or less unbroken. This can only be accomplished by accurate surveys, and by a great multiplication of the points of meteorological registry, already so numerous; and as, moreover, considerable changes in the proportion of forest and of cultivated land, or of dry and wholly or partially submerged surface, will often take place within brief periods, it is highly desirable that the attention of observers, in whose neighborhood the clearing of the soil, or the drainage of lakes and swamps, or other great works of rural improvement, are going on or meditated, should be especially drawn not only to revolutions in atmospheric temperature and precipitation, but to the more easily ascertained and perhaps more important local changes produced by these operations in the temperature and the hygrometric state of the superficial strata of the earth, and in its spontaneous vegetable and animal products.

The rapid extension of railroads, which now everywhere keep pace with, and sometimes even precede, the occupation of new soil for agricultural purposes, furnishes great facilities for enlarging our knowledge of the topography of the territory they traverse, because their cuttings reveal the composition and general structure of surface, and the inclination and elevation of their lines constitute known hypsometrical sections, which give numerous points of departure for the measurement of higher and lower stations, and of course for determining the relief and depression of surface, the slope of the beds of watercourses, and many other not less important questions.

The geological, hydrographical, and topographical surveys, which almost every general and even local government of the civilized world is carrying on, are making yet more important contributions to our stock of geographical and general physical knowledge, and, within a comparatively short space, there will be an accumulation of well established constant and historical facts, from which we can safely reason upon all the relations of action and reaction between man and external nature.

But we are, even now, breaking up the floor and wainscoting and doors and window frames of our dwelling, for fuel to warm our bodies and to seethe our pottage, and the world cannot afford to wait till the slow and sure progress of exact science has taught it a better economy. Many practical lessons have been learned by the common observation of unschooled men; and the teachings of simple experience, on topics where natural philosophy has scarcely yet spoken, are not to be despised.

In these humble pages, which do not in the least aspire to rank among scientific expositions of the laws of nature, I shall attempt to give the most important practical conclusions suggested by the history of man's efforts to replenish the earth and subdue it; and I shall aim to support those conclusions by such facts and illustrations only as address themselves to the understanding of every intelligent reader, and as are to be found recorded in works capable of profitable perusal, or at least consultation, by persons who have not enjoyed a special scientific training.

Theodore Roosevelt
WILDERNESS MAKES MEN NEW

Theodore Roosevelt counted naturalist John Muir and conservationist Gifford Pinchot among his friends. He sympathized with Muir and voted with Pinchot. Roosevelt corresponded with Frederick Jackson Turner and anticipated many features of the frontier thesis in his 1889 The Winning of the West. *It was the wilderness, Roosevelt argued, that would save modern Americans from becoming like an "overcivilized man, who has lost the great fighting, masterful virtues." Manliness and other rugged American virtues, he believed, were the product of fundamental human encounters with wilderness and wild life. This was the "strenuous life." Man won the encounters, but nature was not diminished by losing. The natural world was a bottomless wellspring of virility. Roosevelt regularly toured the national parks, including a 1903 primitive camping trip into the Sierras with John Muir. He strongly advocated preservation for the wilderness and conservation for natural resources. He remained torn between Muir's primitivism and Pinchot's "wise-use conservation." Typically, he approved the Hetch Hetchy Valley water reservoir, but later admitted that any dam which flooded the valley was an act of vandalism. The following selection is taken from "The Natural Resources—Their Wise Use or Their Waste," an address which opened a conference on Conservation of Natural Resources at the White House, May 13, 1908.*

With the rise of peoples from savagery to civilization, and with the consequent growth in the extent and variety of the needs of the average man, there comes a steadily increasing growth of the amount demanded by this average man from the actual resources of the country. Yet, rather curiously, at the same time the average man is apt to lose his realization of this dependence upon nature.

Savages, and very primitive peoples generally, concern themselves only with superficial natural resources; with those which they obtain from the actual surface of the ground. As peoples become a little less primitive, their industries, although in a rude manner, are extended to resources below the surface; then, with what we call civilization and the extension of knowledge, more resources come into use, industries are multiplied, and foresight begins to become a necessary and prominent factor in life. Crops are cultivated; animals are domesticated; and metals are mastered.

From *American Problems*, by Theodore Roosevelt, in *The Works of Theodore Roosevelt* (Charles Scribner's Sons, 1925), Vol. XVIII, pp. 158–160, 163–165.

Every step of the progress of mankind is marked by the discovery and use of natural resources previously unused. Without such progressive knowledge and utilization of natural resources population could not grow, nor industries multiply, nor the hidden wealth of the earth be developed for the benefit of mankind.

From the first beginnings of civilization, on the banks of the Nile and the Euphrates, the industrial progress of the world has gone on slowly, with occasional setbacks, but on the whole steadily, through tens of centuries to the present day. But of late the rapidity of the process has increased at such a rate that more space has been actually covered during the century and a quarter occupied by our national life than during the preceding six thousand years that take us back to the earliest monuments of Egypt, to the earliest cities of the Babylonian plain.

When the founders of this nation met at Independence Hall in Philadelphia the conditions of commerce had not fundamentally changed from what they were when the Phœnician keels first furrowed the lonely waters of the Mediterranean. The differences were those of degree, not of kind, and they were not in all cases even those of degree. Mining was carried on fundamentally as it had been carried on by the Pharaohs in the countries adjacent to the Red Sea.

The wares of the merchants of Boston, of Charleston, like the wares of the merchants of Nineveh and Sidon, if they went by water, were carried by boats propelled by sails or oars; if they went by land they were carried in wagons drawn by beasts of draft or in packs on the backs of beasts of burden. The ships that crossed the high seas were better than the ships that had once crossed the Ægean, but they were of the same type, after all—they were wooden ships propelled by sails; and on land, the roads were not as good as the roads of the Roman Empire, while the service of the posts was probably inferior.

In Washington's time anthracite coal was known only as a useless black stone; and the great fields of bituminous coal were undiscovered. As steam was unknown, the use of coal for power production was undreamed of. Water was practically the only source of power, save the labor of men and animals; and this power was used only in the most primitive fashion. But a few small iron deposits had been found in this country, and the use of iron by our countrymen was very small. Wood was practically the only fuel, and what lumber was sawed was consumed locally, while the forests were regarded chiefly as obstructions to settlement and cultivation.

Such was the degree of progress to which civilized mankind had attained when this nation began its career. It is almost impossible for us in this day to realize how little our Revolutionary ancestors knew of the great store of natural resources whose discovery and use have been such vital factors in the growth and greatness of this nation, and how little they required to take from this store in order to satisfy their needs.

Since then our knowledge and use of the resources of the present territory of the United States have increased a hundredfold. Indeed, the growth of this nation by leaps and bounds makes one of the most striking and important chapters in the history of the world. Its growth has been due to the rapid development, and alas! that it should be said, to the rapid destruction of our natural resources. Nature has supplied to us in the United States, and still supplies to us, more kinds of resources in a more lavish degree than has ever been the case at any other time or with any other people. Our position in the world has been attained by the extent and thoroughness of the control we have achieved over nature; but we are more, and not less, dependent upon what she furnishes than at any previous time of history since the days of primitive man.

*   *   *

We have become great because of the lavish use of our resources and we have just reason to be proud of our growth. But the time has come to inquire seriously what will happen when our forests are gone, when the coal, the iron, the oil, and the gas are exhausted, when the soils shall have been still further impoverished and washed into the streams, polluting the rivers, denuding the fields, and obstructing navigation. These questions do not relate only to the next century or to the next generation. It is time for us now as a nation to exercise the same reasonable foresight in dealing with our great natural resources that would be shown by any prudent man in conserving and widely using the property which contains the assurance of well-being for himself and his children.

The natural resources I have enumerated can be divided into two sharply distinguished classes accordingly as they are or are not capable of renewal. Mines if used must necessarily be exhausted. The minerals do not and cannot renew themselves. Therefore in dealing with the coal, the oil, the gas, the iron, the metals generally, all that we can do is to try to see that they are wisely used. The exhaustion is certain to come in time.

The second class of resources consists of those which cannot only be used in such manner as to leave them undiminished for our children, but can actually be improved by wise use. The soil, the forests, the waterways come in this category. In dealing with mineral resources, man is able to improve on nature only by putting the resources to a beneficial use which in the end exhausts them; but in dealing with the soil and its products man can improve on nature by compelling the resources to renew and even reconstruct themselves in such manner as to serve increasingly beneficial uses— while the living waters can be so controlled as to multiply their benefits.

Neither the primitive man nor the pioneer was aware of any duty to posterity in dealing with the renewable resources. When the American settler felled the forests, he felt that there was plenty of forest left for the sons who came after him. When he exhausted the soil of his farm he felt that his son could go West and take up another. So it was with his immediate successors. When the soil wash from the farmer's fields choked the neighboring river he thought only of using the railway rather than boats for moving his produce and supplies.

Now all this is changed. . . .

*   *   *

. . . Especial credit is due to the initiative, the energy, the devotion to duty, and the far-sightedness of Gifford Pinchot, to whom we owe so much of the progress we have already made in handling this matter of the co-ordination and conservation of natural resources.

John Muir

IN WILDNESS IS THE PRESERVATION
OF THE WORLD

*Scotch-born, Wisconsin-bred, and Sierra-bound, John Muir (1838–1914) called
himself a "poetico-trampo-geologist-botanist-ornithnatural, etc!-!-!-!" Living
in California, Nevada, Utah, the Northwest, and Alaska, Muir made wilderness preservation into a national crusade. Muir's life in the California Sierras,
and his discovery of Romantic Transcendentalists like Wordsworth, Emerson,
and Thoreau made him the major champion for wild nature. At first he was
an eager associate of Gifford Pinchot, but later angrily repudiated Pinchot's
plan to manage American resources primarily to build an industrial civilization. He believed this was a stifling program, for "a little pure wilderness is
the one great present want, both of men and sheep." Three articles moved
public opinion: "Forest Reservations and National Parks," in the June 5,
1897, issue of Harper's Weekly; "The American Forests," in the August 1897
Atlantic Monthly; and "The Wild Parks and Forest Reservations of the West,"
in the same journal in January 1898. He had already founded the Sierra
Club in 1892 and become its first president. He was the guiding force behind
the establishment of Yosemite National Park, largely through articles in the
Century Magazine and Scribner's Monthly in 1889 and 1890. He also fought,
with limited success, to establish a system of national forests, and he re-
sisted plans to make Yosemite's Hetch Hetchy Valley into a reservoir.*

The tendency nowadays to wander in wildernesses is delightful to
see. Thousands of tired, nerve-shaken, over-civilized people are beginning to find out that going to the mountains is going home; that
wildness is a necessity; and that mountain parks and reservations
are useful not only as fountains of timber and irrigating rivers, but
as fountains of life. Awakening from the stupefying effects of the
vice of over-industry and the deadly apathy of luxury, they are trying
as best they can to mix and enrich their own little ongoings with
those of Nature, and to get rid of rust and disease. Briskly venturing
and roaming, some are washing off sins and cobweb cares of the
devil's spinning in all-day storms on mountains; sauntering in rosiny
pinewoods or in gentian meadows, brushing through chaparral,
bending down and parting sweet, flowery sprays; tracing rivers to
their sources, getting in touch with the nerves of Mother Earth;

From "The Wild Parks and Forest Reservations of the West," by John Muir, in the
Atlantic Monthly, Vol. LXXXI, No. 483 (January 1898).

jumping from rock to rock, feeling the life of them, learning the songs of them, panting in whole-souled exercise and rejoicing in deep, long-drawn breaths of pure wildness. This is fine and natural and full of promise. And so also is the growing interest in the care and preservation of forests and wild places in general, and in the half-wild parks and gardens of towns. Even the scenery habit in its most artificial forms, mixed with spectacles, silliness, and kodaks; its devotees arrayed more gorgeously than scarlet tanagers, frightening the wild game with red umbrellas,—even this is encouraging, and may well be regarded as a hopeful sign of the times.

All the Western mountains are still rich in wildness, and by means of good roads are being brought nearer civilization every year. To the sane and free it will hardly seem necessary to cross the continent in search of wild beauty, however easy the way, for they find it in abundance wherever they chance to be. Like Thoreau they see forests in orchards and patches of huckleberry brush, and oceans in ponds and drops of dew. Few in these hot, dim, frictiony times are quite sane or free; choked with care like clocks full of dust, laboriously doing so much good and making so much money,—or so little,—they are no longer good themselves.

When, like a merchant taking a list of his goods, we take stock of our wildness, we are glad to see how much of even the most destructible kind is still unspoiled. Looking at our continent as scenery when it was all wild, lying between beautiful seas, the starry sky above it, the starry rocks beneath it, to compare its sides, the East and the West, would be like comparing the sides of a rainbow. But it is no longer equally beautiful.

*　　*　　*

. . . Man, too, is making many far-reaching changes. This most influential half animal, half angel is rapidly multiplying and spreading, covering the seas and lakes with ships, the land with huts, hotels, cathedrals, and clustered city shops and homes, so that soon, it would seem, we may have to go farther than Nansen to find a good sound solitude. None of Nature's landscapes are ugly so long as they are wild; and much, we can say comfortingly, must always be in great part wild, particularly the sea and the sky, the floods of light from the stars, and the warm, unspoilable heart of the earth, infinitely beautiful, though only dimly visible to the eye of imagination. The geysers, too, spouting from the hot underworld; the steady, long-lasting glaciers on the mountains, obedient only to the sun; Yosemite domes and the tremendous grandeur of rocky cañons and mountains in general,—these must always be wild, for man can change them and mar them hardly more than can the butterflies that hover above them. But the continent's outer beauty is fast passing away, especially the plant part of it, the most destructible and most universally charming of all.

Only thirty years ago, the great Central Valley of California, five hundred miles long and fifty miles wide, was one bed of golden and purple flowers. Now it is ploughed and pastured out of existence, gone forever,—scarce a memory of it left in fence corners and along the bluffs of the streams. The gardens of the Sierra also, and the noble forests in both the reserved and the unreserved portions, are sadly hacked and trampled, notwithstanding the ruggedness of the topography,—all excepting those of the parks guarded by a few soldiers. In the noblest forests of the world, the ground, once divinely beautiful, is desolate and repulsive, like a face ravaged by disease. This is true also of many other Pacific Coast and Rocky Mountain valleys and forests. The same fate, sooner or later, is awaiting them all, unless awakening public opinion comes forward to stop it. Even the great deserts in Arizona, Nevada, Utah, and New Mexico, which offer so little to attract settlers, and which a few years ago pioneers were afraid of, as places of desolation and death, are now taken as pastures at the rate of one or two square miles per cow, and of course their plant treasures are passing away,—the delicate abronias, phloxes, gilias, etc. Only a few of the bitter, thorny, unbitable shrubs are left, and the sturdy cactuses that defend themselves with bayonets and spears.

Most of the wild plant wealth of the East also has vanished,— gone into dusty history. Only vestiges of its glorious prairie and woodland wealth remain to bless humanity in boggy, rocky, unploughable places. Fortunately, some of these are purely wild, and go far to keep Nature's love visible.

*   *   *

. . . The wildest health and pleasure grounds accessible and available to tourists seeking escape from car and dust and early death are the parks and reservations of the West. There are four national parks,—the Yellowstone, Yosemite, General Grant, and Sequoia,—all within easy reach, and thirty forest reservations, a magnificent realm of woods, most of which, by railroads and trails and open ridges, is also fairly accessible, not only to the determined traveler rejoicing in difficulties, but to those (may their tribe increase) who, not tired, not sick, just naturally take wing every summer in search of wildness. The forty million acres of these reserves are in the main unspoiled as yet, though sadly wasted and threatened on their more open margins by the axe and fire of the lumberman and prospector, and by hoofed locusts, which, like the winged ones, devour every leaf within reach, while the shepherds and owners set fires with the intention of making a blade of grass grow in the place of every tree, but with the result of killing both the grass and the trees.

*   *   *

In calm Indian summer, when the heavy winds are hushed, the vast forests covering hill and dale, rising and falling over the rough topography and vanishing in the distance, seem lifeless. No moving thing is seen as we climb the peaks, and only the low, mellow murmur of falling water is heard, which seems to thicken the silence. Nevertheless, how many hearts with warm red blood in them are beating under cover of the woods, and how many teeth and eyes are shining! A multitude of animal people, intimately related to us, but of whose lives we know almost nothing, are as busy about their own affairs as we are about ours: beavers are building and mending dams and huts for winter, and storing them with food; bears are studying winter quarters as they stand thoughtful in open spaces, while the gentle breeze ruffles the long hair on their backs; elk and deer, assembling on the heights, are considering cold pastures where they will be farthest away from the wolves; squirrels and marmots are busily laying up provisions and lining their nests against coming frost and snow foreseen; and countless thousands of birds are forming parties and gathering their young about them for flight to the southlands; while butterflies and bees, apparently with no thought of hard times to come, are hovering above the late-blooming goldenrods, and, with countless other insect folk, are dancing and humming right merrily in the sunbeams and shaking all the air into music.

Wander here a whole summer, if you can. Thousands of God's wild blessings will search you and soak you as if you were a sponge, and the big days will go by uncounted. But if you are business-tangled, and so burdened with duty that only weeks can be got out of the heavy-laden year, then go to the Flathead Reserve; for it is easily and quickly reached by the Great Northern Railroad. Get off the track at Belton Station, and in a few minutes you will find your-self in the midst of what you are sure to say is the best care-killing scenery on the continent,—beautiful lakes derived straight from glaciers, lofty mountains steeped in lovely nemophila-blue skies and clad with forests and glaciers, mossy, ferny waterfalls in their hol-lows, nameless and numberless, and meadowy gardens abounding in the best of everything.

*   *   *

Notwithstanding the outcry against the reservations last winter in Washington, that uncounted farms, towns, and villages were in-cluded in them, and that all business was threatened or blocked, nearly all the mountains in which the reserves lie are still covered with virgin forests. Though lumbering has long been carried on with tremendous energy along their boundaries, and home-seekers have explored the woods for openings available for farms, however small, one may wander in the heart of the reserves for weeks without meeting a human being, Indian or white man, or any conspicuous trace of one. Indians used to ascend the main streams on their way to the mountains for wild goats, whose wool furnished them clothing. But with food in abundance on the coast there was little to draw them into the woods, and the monuments they have left there are scarcely more conspicuous than those of birds and squirrels; far less so than those of the beavers, which have dammed streams and made clearings that will endure for centuries. Nor is there much in these woods to attract cattle-keepers. Some of the first settlers made farms on the small bits of prairie and in the comparatively open Cowlitz and Chehalis valleys of Washington; but before the gold period most of the immigrants from the Eastern States settled in the fertile and open Willamette Valley of Oregon. Even now, when the search for tillable land is so keen, excepting the bottom-lands of the rivers around Puget Sound, there are few cleared spots in all western Washington. On every meadow or opening of any sort some one will be found keeping cattle, raising hops, or cultivating patches of grain, but these spots are few and far between. All the larger spaces were taken long ago; therefore most of the newcomers build their cabins where the beavers built theirs. They keep a few cows, laboriously widen their little meadow openings by hacking, girdling, and burning the rim of the close-pressing forest, and scratch and plant among the huge blackened logs and stumps, girdling and killing themselves in killing the trees.

*   *   *

These grand reservations should draw thousands of admiring visitors at least in summer, yet they are neglected as if of no account, and spoilers are allowed to ruin them as fast as they like. A few peeled spars cut here were set up in London, Philadelphia, and Chicago, where they excited wondering attention; but the countless hosts of living trees rejoicing at home on the mountains are scarce considered at all. Most travelers here are content with what they can see from car windows or the verandas of hotels, and in going from place to place cling to their precious trains and stages like wrecked sailors to rafts. When an excursion into the woods is proposed, all sorts of dangers are imagined,—snakes, bears, Indians. Yet it is far safer to wander in God's woods than to travel on black highways or to stay at home. The snake danger is so slight it is hardly worth mentioning. Bears are a peaceable people, and mind their own business, instead of going about like the devil seeking whom they may devour. Poor fellows, they have been poisoned, trapped, and shot at until they have lost confidence in brother man, and it is not now easy to make their acquaintance. As to Indians, most of them are dead or civilized into useless innocence. No American wilderness that I know of is so dangerous as a city home "with all the modern improvements." One should go to the woods for safety, if for nothing else. Lewis and Clark, in their famous trip across the continent in 1804–1805, did not lose a single man by Indians or animals, though all the West was then wild. Captain Clark was bitten on the hand as he lay asleep. That was one bite among more than a hundred men while traveling nine thousand miles. Loggers are far more likely to be met than Indians or bears in the reserves or about their boundaries, brown weather-tanned men with faces furrowed like bark, tired-looking, moving slowly, swaying like the trees they chop. A little of everything in the woods is fastened to their clothing, rosiny and smeared with balsam, and rubbed into it, so that their scanty outer garments grow thicker with use and never wear out. Many a forest giant have these old woodmen felled, but, round-shouldered and stooping, they too are leaning over and tottering to their fall. Others, however, stand ready to take their places, stout young fellows, erect as saplings; and always the foes of trees outnumber their friends. Far up the white peaks one can hardly fail to meet the wild goat, or American chamois,—an admirable mountaineer, familiar with woods and glaciers as well as rocks, —and in leafy thickets deer will be found; while gliding about unseen there are many sleek furred animals enjoying their beautiful lives, and birds also, notwithstanding few are noticed in hasty walks. The ousel sweetens the glens and gorges where the streams flow fastest, and every grove has its singers, however silent it seems,—thrushes, linnets, warblers; humming-birds glint about the fringing bloom of the meadows and peaks, and the lakes are stirred into lively pictures by water-fowl.

*   *   *

This Sierra Reserve, proclaimed by the President of the United States in September, 1893, is worth the most thoughtful care of the government for its own sake, without considering its value as the fountain of the rivers on which the fertility of the great San Joaquin Valley depends. Yet it gets no care at all. In the fog of tariff, silver, and annexation politics it is left wholly unguarded, though the management of the adjacent national parks by a few soldiers shows how well and how easily it can be preserved. In the meantime, lumbermen are allowed to spoil it at their will, and sheep in uncountable ravenous hordes to trample it and devour every green leaf within reach; while the shepherds, like destroying angels, set innumerable fires, which burn not only the undergrowth of seedlings on which the permanence of the forest depends, but countless thousands of the venerable giants. If every citizen could take one walk through this reserve, there would be no more trouble about its care; for only in darkness does vandalism flourish.

*   *   *

The Grand Cañon Reserve of Arizona, of nearly two million acres, or the most interesting part of it, as well as the Rainier region, should be made into a national park, on account of their supreme grandeur and beauty. Setting out from Flagstaff, a station on the Atchison, Topeka, and Santa Fé Railroad, on the way to the cañon you pass through beautiful forests of yellow pine,—like those of the Black Hills, but more extensive,—and curious dwarf forests of nut pine and juniper, the spaces between the miniature trees planted with many interesting species of eriogonum, yucca, and cactus. After riding or walking seventy-five miles through these pleasure-grounds, the San Francisco and other mountains, abounding in flowery parklike openings and smooth shallow valleys with long vistas which in fineness of finish and arrangement suggest the work of a consummate landscape artist, watching you all the way, you come to the most tremendous cañon in the world. It is abruptly countersunk in the forest plateau, so that you see nothing of it until you are suddenly stopped on its brink, with its immeasurable wealth of divinely colored and sculptured buildings before you and beneath you. No matter how far you have wandered hitherto, or how many famous gorges and valleys you have seen, this one, the Grand Cañon of the Colorado, will seem as novel to you, as unearthly in the color and grandeur and quantity of its architecture, as if you had found it after death, on some other star; so incomparably lovely and grand and supreme is it above all the other cañons in our fire-moulded, earthquake-shaken, rain-washed, wave-washed, river and glacier sculptured world. It is about six thousand feed deep where you first see it, and from rim to rim ten to fifteen miles wide. Instead of being dependent for interest upon waterfalls, depth, wall sculpture, and beauty of parklike floor, like most other great cañons, it has no waterfalls in sight, and no appreciable floor spaces. The big river has just room enough to flow and roar obscurely, here and there groping its way as best it can, like a weary, murmuring, overladen traveler trying to escape from the tremendous, bewildering labyrinthic abyss, while its roar serves only to deepen the silence. Instead of being filled with air, the vast space between the walls is crowded with Nature's grandest buildings,—a sublime city of them, painted in every color, and adorned with richly fretted cornice and battlement spire and tower in endless variety of style and architecture. Every architectural invention of man has been anticipated, and far more, in this grandest of God's terrestrial cities.

Gifford Pinchot
WILD LAND IS WASTED LAND

Gifford Pinchot (1865–1948), the first professional American forester, was the major early advocate of the "wise use" school of conservation. Trained in European land management for "maximum sustained yield," Pinchot developed forestry practices he believed were required to fulfill the material needs of a growing nation without destroying future aesthetic, spiritual, and practical use. The close affinity between Pinchot and John Muir in their early relationships became strained over whether preservation or conservation would be the policy of the new Forestry Commission in the 1890's. As Roderick Nash has noted, "For all his love of the woods, Pinchot's ultimate loyalty was to civilization and forestry; Muir's to wilderness and preservation." The existence of pure wilderness was incompatible with forest management. Pinchot became chief of the Forest Service and played an important role as policy-maker on numerous federal waterways and conservation commissions. His point of view still dominates today's federal land-use programs, although the ecology controversy in the 1960's has encouraged the "wilderness-for-wilderness-sake" preservationists. The following selection is taken from The Fight for Conservation, *Pinchot's mature progressive confession of faith and plea for conservation.*

The first great fact about conservation is that it stands for development. There has been a fundamental misconception that conservation means nothing but the husbanding of resources for future generations. There could be no more serious mistake. Conservation does mean provision for the future, but it means also and first of all the recognition of the right of the present generation to the fullest

From *The Fight for Conservation,* by Gifford Pinchot (Doubleday, Page and Company, 1910).

necessary use of all the resources with which this country is so abundantly blessed. Conservation demands the welfare of this generation first, and afterward the welfare of the generations to follow.

The first principle of conservation is development, the use of the natural resources now existing on this continent for the benefit of the people who live here now. There may be just as much waste in neglecting the development and use of certain natural resources as there is in their destruction. We have a limited supply of coal, and only a limited supply. Whether it is to last for a hundred or a hundred and fifty or a thousand years, the coal is limited in amount, unless through geological changes which we shall not live to see, there will never be any more of it than there is now. But coal is in a sense the vital essence of our civilization. If it can be preserved, if the life of the mines can be extended, if by preventing waste there can be more coal left in this country after we of this generation have made every needed use of this source of power, then we shall have deserved well of our descendants.

Conservation stands emphatically for the development and use of water-power now, without delay. It stands for the immediate construction of navigable waterways under a broad and comprehensive plan as assistants to the railroads. More coal and more iron are required to move a ton of freight by rail than by water, three to one. In every case and in every direction the conservation movement has development for its first principle, and at the very beginning of its work. The development of our natural resources and the fullest use of them for the present generation is the first duty of this generation. So much for development.

In the second place conservation stands for the prevention of waste. There has come gradually in this country an understanding that waste is not a good thing and that the attack on waste is an industrial necessity. I recall very well indeed how, in the early days of forest fires, they were considered simply and solely as acts of God, against which any opposition was hopeless and any attempt to control them not merely hopeless but childish. It was assumed that they came in the natural order of things, as inevitably as the seasons or the rising and setting of the sun. To-day we understand that forest fires are wholly within the control of men. So we are coming in like manner to understand that the prevention of waste in all other directions is a simple matter of good business. The first duty of the human race is to control the earth it lives upon.

We are in a position more and more completely to say how far the waste and destruction of natural resources are to be allowed to go on and where they are to stop. It is curious that the effort to stop waste, like the effort to stop forest fires, has often been considered as a matter controlled wholly by economic law. I think there could be no greater mistake. Forest fires were allowed to burn long after the people had means to stop them. The idea that men were helpless in the face of them held long after the time had passed when the means of control were fully within our reach. It was the old story that "as a man thinketh, so is he"; we came to see that we could stop forest fires, and we found that the means had long been at hand. When at length we came to see that the control of logging in certain directions was profitable, we found it had long been possible. In all these matters of waste of natural resources, the education of the people to understand that they can stop the leakage comes before the actual stopping and after the means of stopping it have long been ready at our hands.

In addition to the principles of development and preservation of our resources there is a third principle. It is this: The natural resources must be developed and preserved for the benefit of the many, and not merely for the profit of a few. We are coming to understand in this country that public action for public benefit has a very much wider field to cover and a much larger part to play than was the case when there were resources enough for every one, and before certain constitutional provisions had given so tremendously strong a position to vested rights and property in general.

*　　　*　　　*

Danger to a nation comes either from without or from within. In the first great crisis of our history, the Revolution, another people attempting from without to halt the march of our destiny by refusing to us liberty. With reasonable prudence and preparedness we need never fear another such attempt. If there be danger, it is not from an external source. In the second great crisis, the Civil War, a part of our own people strove for an end which would have checked the progress of development. Another such attempt has become forever impossible. If there be danger, it is not from a division of our people.

In the third great crisis of our history, which has now come squarely upon us, the special interests and the thoughtless citizens seem to have united together to deprive the Nation of the great natural resources without which it cannot endure. This is the pressing danger now, and it is not the least to which our National life has been exposed. A nation deprived of liberty may win it, a nation divided may reunite, but a nation whose natural resources are destroyed must inevitably pay the penalty of poverty, degradation, and decay.

At first blush this may seem like an unpardonable misconception and over-statement, and if it is not true it certainly is unpardonable. Let us consider the facts. Some of them are well known, and the salient ones can be put very briefly.

The five indispensably essential materials in our civilization are wood, water, coal, iron, and agricultural products.

We have timber for less than thirty years at the present rate of cutting. The figures indicate that our demands upon the forest have increased twice as fast as our population.

We have anthracite coal for but fifty years, and bituminous coal for less than two hundred.

Our supplies of iron ore, mineral oil, and natural gas are being rapidly depleted, and many of the great fields are already exhausted. Mineral resources such as those when once gone are gone forever.

We have allowed erosion, that great enemy of agriculture, to impoverish and, over thousands of square miles, to destroy our farms.

*   *   *

We have a well-marked national tendency to disregard the future, and it has led us to look upon all our natural resources as inexhaustible. Even now that the actual exhaustion of some of them is forcing itself upon us in higher prices and the greater cost of living, we are still asserting, if not always in words, yet in the far stronger language of action, that nevertheless and in spite of it all, they still are inexhaustible.

It is this national attitude of exclusive attention to the present, this absence of foresight from among the springs of national action, which is directly responsible for the present condition of our natural resources. It was precisely the same attitude which brought Palestine, once rich and populous, to its present desert condition, and which destroyed the fertility and habitability of vast areas in northern Africa and elsewhere in so many of the older regions of the world.

The conservation of our natural resources is a question of primary importance on the economic side. It pays better to conserve our natural resources than to destroy them, and this is especially true when the national interest is considered. But the business reason, weighty and worthy though it be, is not the fundamental reason. In such matters, business is a poor master but a good servant. The law of self-preservation is higher than the law of business, and the duty of preserving the Nation is still higher than either.

*   *   *

We have passed the inevitable stage of pioneer pillage of natural resources. The natural wealth we found upon this continent has made us rich. We have used it, as we had a right to do, but we have not stopped there. We have abused, and wasted, and exhausted it also, so that there is the gravest danger that our prosperity to-day will have been bought at the price of the suffering and poverty of our descendants. We may now fairly ask of ourselves a reasonable care for the future and a natural interest in those who are to come after us. No patriotic citizen expects this Nation to run its course and perish in a hundred or two hundred, or five hundred years; but, on the contrary, we expect it to grow in influence and power and, what is of vastly greater importance, in the happiness and prosperity of our people. But we have as little reason to expect that all this will happen of itself as there would have been for the men who established this Nation to expect that a United States would grow of itself without their efforts and sacrifices. It was their duty to found this Nation, and they did it. It is our duty to provide for its continuance in well-being and honor. That duty it seems as though we might neglect—not in wilfulness, not in any lack of patriotic devotion, when once our patriotism is aroused, but in mere thoughtlessness and inability or unwillingness to drop the interests of the moment long enough to realize that what we do now will decide the future of the Nation. For, if we do not take action to conserve the Nation's natural resources, and that soon, our descendants will suffer the penalty of our neglect.

Aldo Leopold

IMMORAL MAN AND THE MORAL UNIVERSE

Aldo Leopold was the first modern naturalist who developed a poetic and ethic of ecology. One of the first professional naturalists, he was a member of the United States Forest Service, and concentrated on wildlife management and wilderness preservation. While advocates of wilderness like Emerson and Muir sought to arouse a mystical intimacy with nature, Leopold created a philosophical, religious and ethical point of view based on pragmatic scientific grounds. Leopold believed that man's well-being was more than a case of better economic development or political achievement. Human life was an act of celebration. This celebration gained reality when man realized he belonged to the environment and shared the dynamic of life with all living things. Man's unique powers to change and manipulate the environment made him uniquely responsible to maintain and perserve it as the center of celebration and life. Leopold helped to pioneer ecology as a science—and as a comprehensive system of values—by understanding the environment as a complex series of interrelated organisms. He died in 1948 while fighting a brush fire along the Wisconsin River.

When god-like Odysseus returned from the wars in Troy, he hanged all on one rope a dozen slave-girls of his household whom he suspected of misbehavior during his absence.

This hanging involved no question of propriety. The girls were property. The disposal of property was then, as now, a matter of expediency, not of right and wrong.

Concepts of right and wrong were not lacking from Odysseus'

From *A Sand County Almanac* by Aldo Leopold. Copyright © 1949, 1966 by Oxford University Press, Inc. Reprinted by permission.

Greece: witness the fidelity of his wife through the long years before at last his black-prowed galleys clove the wine-dark seas for home. The ethical structure of that day covered wives, but had not yet been extended to human chattels. During the three thousand years which have since elapsed, ethical criteria have been extended to many fields of conduct, with corresponding shrinkages in those judged by expediency only.

The Ethical Sequence

This extension of ethics, so far studied only by philosophers, is actually a process in ecological evolution. Its sequences may be described in ecological as well as in philosophical terms. An ethic, ecologically, is a limitation on freedom of action in the struggle for existence. An ethic, philosophically, is a differentiation of social from anti-social conduct. These are two definitions of one thing. The thing has its origin in the tendency of interdependent individuals or groups to evolve modes of co-operation. The ecologist calls these symbioses. Politics and economics are advanced symbioses in which the original free-for-all competition has been replaced, in part, by co-operative mechanisms with an ethical content.

The complexity of co-operative mechanisms has increased with population density, and with the efficiency of tools. It was simpler, for example, to define the anti-social uses of sticks and stones in the days of the mastodons than of bullets and billboards in the age of motors.

The first ethics dealt with the relation between individuals; the Mosaic Decalogue is an example. Later accretions dealt with the relation between the individual and society. The Golden Rule tries to integrate the individual to society; democracy to integrate social organization to the individual.

There is as yet no ethic dealing with man's relation to land and to the animals and plants which grow upon it. Land, like Odysseus' slave-girls, is still property. The land-relation is still strictly economic, entailing privileges but not obligations.

The extension of ethics to this third element in human environment is, if I read the evidence correctly, an evolutionary possibility and an ecological necessity. It is the third step in a sequence. The first two have already been taken. Individual thinkers since the days of Ezekiel and Isaiah have asserted that the despoliation of land is not only inexpedient but wrong. Society, however, has not yet affirmed their belief. I regard the present conservation movement as the embryo of such an affirmation.

An ethic may be regarded as a mode of guidance for meeting ecological situations so new or intricate, or involving such deferred reactions, that the path of social expediency is not discernible to the average individual. Animal instincts are modes of guidance for the individual in meeting such situations. Ethics are possibly a kind of community instinct in-the-making.

The Community Concept

All ethics so far evolved rest upon a single premise: that the individual is a member of a community of interdependent parts. His instincts prompt him to compete for his place in that community, but his ethics prompt him also to co-operate (perhaps in order that there may be a place to compete for).

The land ethic simply enlarges the boundaries of the community to include soils, waters, plants, and animals, or collectively: the land.

This sounds simple: do we not already sing our love for and obligation to the land of the free and the home of the brave? Yes, but just what and whom do we love? Certainly not the soil, which we are sending helter-skelter downriver. Certainly not the waters, which we assume have no function except to turn turbines, float barges, and carry off sewage. Certainly not the plants, of which we exterminate whole communities without batting an eye. Certainly not the animals, of which we have already extirpated many of the largest and most beautiful species. A land ethic of course cannot prevent the alteration, management, and use of these "resources," but it does affirm their right to continued existence, and, at least in spots, their continued existence in a natural state.

In short, a land ethic changes the role of *Homo sapiens* from conqueror of the land-community to plain member and citizen of it. It implies respect for his fellow-members, and also respect for the community as such.

In human history, we have learned (I hope) that the conqueror role is eventually self-defeating. Why? Because it is implicit in such a role that the conqueror knows, *ex cathedra,* just what makes the community clock tick, and just what and who is valuable, and what and who is worthless, in community life. It always turns out that he knows neither, and this is why his conquests eventually defeat themselves.

In the biotic community, a parallel situation exists. Abraham knew exactly what the land was for: it was to drip milk and honey into Abraham's mouth. At the present moment, the assurance with which we regard this assumption is inverse to the degree of our education.

The ordinary citizen today assumes that science knows what makes the community clock tick; the scientist is equally sure that he does not. He knows that the biotic mechanism is so complex that its workings may never be fully understood.

That man is, in fact, only a member of a biotic team is shown by an ecological interpretation of history. Many historical events, hitherto explained solely in terms of human enterprise, were actually biotic interactions between people and land. The characteristics of the land determined the facts quite as potently as the characteristics of the men who lived on it.

Consider, for example, the settlement of the Mississippi valley. In the years following the Revolution, three groups were contending for its control: the native Indian, the French and English traders, and the American settlers. Historians wonder what would have happened if the English at Detroit had thrown a little more weight into the Indian side of those tipsy scales which decided the outcome of the colonial migration into the cane-lands of Kentucky. It is time now to ponder the fact that the cane-lands, when subjected to the particular mixture of forces represented by the cow, plow, fire, and axe of the pioneer, became bluegrass. What if the plant succession inherent in this dark and bloody ground had, under the impact of these forces, given us some worthless sedge, shrub, or weed? Would Boone and Kenton have held out? Would there have been any overflow into Ohio, Indiana, Illinois, and Missouri? Any Louisiana Purchase? Any transcontinental union of new states? Any Civil War?

Kentucky was one sentence in the drama of history. We are

commonly told what the human actors in this drama tried to do, but we are seldom told that their success, or the lack of it, hung in large degree on the reaction of particular soils to the impact of the particular forces exerted by their occupancy. In the case of Kentucky, we do not even know where the bluegrass came from— whether it is a native species, or a stowaway from Europe.

Contrast the cane-lands with what hindsight tells us about the Southwest, where the pioneers were equally brave, resourceful, and persevering. The impact of occupancy here brought no bluegrass, or other plant fitted to withstand the bumps and buffetings of hard use. This region, when grazed by livestock, reverted through a series of more and more worthless grasses, shrubs, and weeds to a condition of unstable equilibrium. Each recession of plant types bred erosion; each increment to erosion bred a further recession of plants. The result today is a progressive and mutual deterioration, not only of plants and soils, but of the animal community subsisting thereon. The early settlers did not expect this: on the ciénegas of New Mexico some even cut ditches to hasten it. So subtle has been its progress that few residents of the region are aware of it. It is quite invisible to the tourist who finds this wrecked landscape colorful and charming (as indeed it is, but it bears scant resemblance to what it was in 1848).

This same landscape was "developed" once before, but with quite different results. The Pueblo Indians settled the Southwest in pre-Columbian times, but they happened *not* to be equipped with range livestock. Their civilization expired, but not because their land expired.

In India, regions devoid of any sod-forming grass have been settled, apparently without wrecking the land, by the simple expedient of carrying the grass to the cow, rather than vice versa. (Was this the result of some deep wisdom, or was it just good luck? I do not know.)

In short, the plant succession steered the course of history; the pioneer simply demonstrated, for good or ill, what successions inhered in the land. Is history taught in this spirit? It will be, once the concept of land as a community really penetrates our intellectual life.

The Ecological Conscience

Conservation is a state of harmony between men and land. Despite nearly a century of propaganda, conservation still proceeds at a snail's pace; progress still consists largely of letterhead pieties and convention oratory. On the back forty we still slip two steps backward for each forward stride.

The usual answer to this dilemma is "more conservation education." No one will debate this, but is it certain that only the *volume* of education needs stepping up? Is something lacking in the *content* as well?

It is difficult to give a fair summary of its content in brief form, but, as I understand it, the content is substantially this: obey the law, vote right, join some organizations, and practice what conservation is profitable on your own land; the government will do the rest.

Is not this formula too easy to accomplish anything worth-while? It defines no right or wrong, assigns no obligation, calls for no sacrifice, implies no change in the current philosophy of values. In respect of land-use, it urges only enlightened self-interest. Just how far will such education take us? An example will perhaps yield a partial answer.

By 1930 it had become clear to all except the ecologically blind that southwestern Wisconsin's topsoil was slipping seaward. In 1933 the farmers were told that if they would adopt certain remedial practices for five years, the public would donate CCC labor to install them, plus the necessary machinery and materials. The offer was widely accepted, but the practices were widely forgotten when the five-year contract period was up. The farmers continued only those practices that yielded an immediate and visible economic gain for themselves.

This led to the idea that maybe farmers would learn more quickly if they themselves wrote the rules. Accordingly the Wisconsin Legislature in 1937 passed the Soil Conservation District Law. This said to farmers, in effect: *We, the public, will furnish you free technical service and loan you specialized machinery, if you will write your own rules for land-use. Each county may write its own rules, and these will have the force of law.* Nearly all the counties promptly organized to accept the proffered help, but after a decade of operation, *no country has yet written a single rule.* There has been visible progress in such practices as strip-cropping, pasture renovation, and soil liming, but none in fencing woodlots against grazing, and none in excluding plow and cow from steep slopes. The farmers, in short, have selected those remedial practices which were profitable anyhow, and ignored those which were profitable to the community, but not clearly profitable to themselves.

When one asks why no rules have been written, one is told that the community is not yet ready to support them; education must precede rules. But the education actually in progress makes no mention of obligations to land over and above those dictated by self-interest. The net result is that we have more education but less soil, fewer healthy woods, and as many floods as in 1937.

The puzzling aspect of such situations is that the existence of obligations over and above self-interest is taken for granted in such rural community enterprises as the betterment of roads, schools, churches, and baseball teams. Their existence is not taken for granted, nor as yet seriously discussed, in bettering the behavior of the water that falls on the land, or in the preserving of the beauty or diversity of the farm landscape. Land-use ethics are still governed wholly by economic self-interest, just as social ethics were a century ago.

To sum up: we asked the farmer to do what he conveniently could to save his soil, and he has done just that, and only that. The farmer who clears the woods off a 75 percent slope, turns his cows into the clearing, and dumps its rainfall, rocks, and soil into the community creek, is still (if otherwise decent) a respected member of society. If he puts lime on his fields and plants his crops on contour, he is still entitled to all the privileges and emoluments of his Soil Conservation District. The District is a beautiful piece of social machinery, but it is coughing along on two cylinders because we have been too timid, and too anxious for quick success, to tell the farmer the true magnitude of his obligations. Obligations have no meaning without conscience, and the problem we face is the extension of the social conscience from people to land.

No important change in ethics was ever accomplished without an internal change in our intellectual emphasis, loyalties, affections, and convictions. The proof that conservation has not yet touched these foundations of conduct lies in the fact that philosophy and religion have not yet heard of it. In our attempt to make conservation easy, we have made it trivial.

Substitutes for a Land Ethic

When the logic of history hungers for bread and we hand out a stone, we are at pains to explain how much the stone resembles bread. I now describe some of the stones which serve in lieu of a land ethic.

One basic weakness in a conservation system based wholly on economic motives is that most members of the land community have no economic value. Wildflowers and songbirds are examples. Of the 22,000 higher plants and animals native to Wisconsin, it is doubtful whether more than 5 percent can be sold, fed, eaten, or otherwise put to economic use. Yet these creatures are members of the biotic community, and if (as I believe) its stability depends on its integrity, they are entitled to continuance.

When one of these non-economic categories is threatened, and if we happen to love it, we invent subterfuges to give it economic importance. At the beginning of the century songbirds were supposed to be disappearing. Ornithologists jumped to the rescue with some distinctly shaky evidence to the effect that insects would eat us up if birds failed to control them. The evidence had to be economic in order to be valid.

It is painful to read these circumlocutions today. We have no land ethic yet, but we have at least drawn nearer the point of admitting that birds should continue as a matter of biotic right, regardless of the presence or absence of economic advantage to us.

A parallel situation exists in respect of predatory mammals, raptorial birds, and fish-eating birds. Time was when biologists somewhat overworked the evidence that these creatures preserve the health of game by killing weaklings, or that they control rodents for the farmer, or that they prey only on "worthless" species. Here again, the evidence had to be economic in order to be valid. It is only in recent years that we hear the more honest argument that predators are members of the community, and that no special interest has the right to exterminate them for the sake of a benefit, real or fancied, to itself. Unfortunately this enlightened view is still in the talk stage. In the field the extermination of predators goes merrily on: witness the impending erasure of the timber wolf by fiat of Congress, the Conservation Bureaus, and many state legislatures.

Some species of trees have been "read out of the party" by economics-minded foresters because they grow too slowly, or have too low a sale value to pay as timber crops: white cedar, tamarack, cypress, beech, and hemlock are examples. In Europe, where forestry is ecologically more advanced, the non-commercial tree species are recognized as members of the native forest community, to be preserved as such, within reason. Moreover some (like beech) have been found to have a valuable function in building up soil fertility. The interdependence of the forest and its constituent tree species, ground flora, and fauna is taken for granted.

Lack of economic value is sometimes a character not only of species or groups, but of entire biotic communities: marshes, bogs, dunes, and "deserts" are examples. Our formula in such cases is to relegate their conservation to government as refuges, monuments, or parks. The difficulty is that these communities are usually interspersed with more valuable private lands; the government cannot possibly own or control such scattered parcels. The net effect is that we have relegated some of them to ultimate extinction over large areas. If the private owner were ecologically minded, he would be proud to be the custodian of a reasonable proportion of such areas, which add diversity and beauty to his farm and to his community.

In some instances, the assumed lack of profit in these "waste" areas has proved to be wrong, but only after most of them had been done away with. The present scramble to reflood muskrat marshes is a case in point.

There is a clear tendency in American conservation to relegate to government all necessary jobs that private landowners fail to perform. Government ownership, operation, subsidy, or regulation is now widely prevalent in forestry, range management, soil and watershed management, park and wilderness conservation, fisheries management, and migratory bird management, with more to come. Most of this growth in governmental conservation is proper and logical, some of it is inevitable. That I imply no disapproval of it is implicit in the fact that I have spent most of my life working for it. Nevertheless the question arises: What is the ultimate magnitude of the enterprise? Will the tax base carry its eventual ramifications? At what point will governmental conservation, like the mastodon, become handicapped by its own dimensions? The answer, if there is any, seems to be in a land ethic, or some other force which assigns more obligation to the private landowner.

Industrial landowners and users, especially lumbermen and stockmen, are inclined to wail long and loudly about the extension of government ownership and regulation to land, but (with notable exceptions) they show little disposition to develop the only visible alternative: the voluntary practice of conservation on their own lands.

When the private landowner is asked to perform some unprofitable act for the good of the community, he today assents only with outstretched palm. If the act costs him cash this is fair and proper, but when it costs only forethought, open-mindedness, or time, the issue is at least debatable. The overwhelming growth of land-use subsidies in recent years must be ascribed, in large part, to the government's own agencies for conservation education: the land bureaus, the agricultural colleges, and the extension services. As far as I can detect, no ethical obligation toward land is taught in these institutions.

To sum up: a system of conservation based solely on economic self-interest is hopelessly lopsided. It tends to ignore, and thus eventually to eliminate, many elements in the land community that lack commercial value, but that are (as far as we know) essential to its healthy functioning. It assumes, falsely, I think, that the economic parts of the biotic clock will function without the uneconomic parts. It tends to relegate to government many functions eventually too large, too complex, or too widely dispersed to be performed by government.

An ethical obligation on the part of the private owner is the only visible remedy for these situations.

The Land Pyramid

An ethic to supplement and guide the economic relation to land presupposes the existence of some mental image of land as a biotic mechanism. We can be ethical only in relation to something we can see, feel, understand, love, or otherwise have faith in.

The image commonly employed in conservation education is "the balance of nature." For reasons too lengthy to detail here, this figure of speech fails to describe accurately what little we know about the land mechanism. A much truer image is the one employed in ecology: the biotic pyramid. I shall first sketch the pyramid as a symbol of land, and later develop some of its implications in terms of land-use.

Plants absorb energy from the sun. This energy flows through a circuit called the biota, which may be represented by a pyramid consisting of layers. The bottom layer is the soil. A plant layer rests on the soil, an insect layer on the plants, a bird and rodent layer on the insects, and so on up through various animal groups to the apex layer, which consists of the larger carnivores.

The species of a layer are alike not in where they came from, or in what they look like, but rather in what they eat. Each successive layer depends on those below it for food and often for other services, and each in turn furnishes food and services to those above. Proceeding upward, each successive layer decreases in numerical abundance. Thus, for every carnivore there are hundreds of his prey, thousands of their prey, millions of insects, uncountable plants. The pyramidal form of the system reflects this numerical progression from apex to base. Man shares an intermediate layer with the bears, raccoons, and squirrels which eat both meat and vegetables.

The lines of dependency for food and other services are called food chains. Thus soil-oak-deer-Indian is a chain that has now been largely converted to soil-corn-cow-farmer. Each species, including ourselves, is a link in many chains. The deer eats a hundred plants other than oak, and the cow a hundred plants other than corn. Both, then, are links in a hundred chains. The pyramid is a tangle of chains so complex as to seem disorderly, yet the stability of the system proves it to be a highly organized structure. Its functioning depends on the co-operation and competition of its diverse parts.

In the beginning, the pyramid of life was low and squat; the food chains short and simple. Evolution has added layer after layer, link after link. Man is one of thousands of accretions to the height and complexity of the pyramid. Science has given us many doubts, but it has given us at least one certainty: the trend of evolution is to elaborate and diversify the biota.

Land, then, is not merely soil; it is a fountain of energy flowing through a circuit of soils, plants, and animals. Food chains are the living channels which conduct energy upward; death and decay return it to the soil. The circuit is not closed; some energy is dissipated in decay, some is added by absorption from the air, some is stored in soils, peats, and long-lived forests; but it is a sustained circuit, like a slowly augmented revolving fund of life. There is always a net loss by downhill wash, but this is normally small and offset by the decay of rocks. It is deposited in the ocean and, in the course of geological time, raised to form new lands and new pyramids.

The velocity and character of the upward flow of energy depend on the complex structure of the plant and animal community, much as the upward flow of sap in a tree depends on its complex cellular organization. Without this complexity, normal circulation would presumably not occur. Structure means the characteristic numbers, as well as the characteristic kinds and functions, of the component species. This interdependence between the complex structure of the land and its smooth functioning as an energy unit is one of its basic attributes.

When a change occurs in one part of the circuit, many other parts must adjust themselves to it. Change does not necessarily obstruct or divert the flow of energy; evolution is a long series of self-induced changes, the net result of which has been to elaborate the flow mechanism and to lengthen the circuit. Evolutionary changes, however, are usually slow and local. Man's invention of tools has enabled him to make changes of unprecedented violence, rapidity, and scope.

One change is in the composition of floras and faunas. The larger predators are lopped off the apex of the pyramid; food chains, for the first time in history, become shorter rather than longer. Domesticated species from other lands are substituted for wild ones, and wild ones are moved to new habitats. In this world-wide pooling of faunas and floras, some species get out of bounds as pests and diseases, others are extinguished. Such effects are seldom intended or foreseen; they represent unpredicted and often untraceable readjustments in the structure. Agricultural science is largely a race between the emergence of new pests and the emergence of new techniques for their control.

Another change touches the flow of energy through plants and animals and its return to the soil. Fertility is the ability of soil to receive, store, and release energy. Agriculture, by overdrafts on the soil, or by too radical a substitution of domestic for native species in the superstructure, may derange the channels of flow or deplete storage. Soils depleted of their storage, or of the organic matter which anchors it, wash away faster than they form. This is erosion.

Waters, like soil, are part of the energy circuit. Industry, by polluting waters or obstructing them with dams, may exclude the plants and animals necessary to keep energy in circulation.

Transportation brings about another basic change: the plants or animals grown in one region are now consumed and returned to the soil in another. Transportation taps the energy stored in rocks, and in the air, and uses it elsewhere; thus we fertilize the garden with nitrogen gleaned by the guano birds from the fishes of seas on the other side of the Equator. Thus the formerly localized and self-contained circuits are pooled on a world-wide scale.

The process of altering the pyramid for human occupation releases stored energy, and this often gives rise, during the pioneering period, to a deceptive exuberance of plant and animal life, both wild and tame. These releases of biotic capital tend to becloud or postpone the penalties of violence.

This thumbnail sketch of land as an energy circuit conveys three basic ideas:

1. That land is not merely soil.
2. That the native plants and animals kept the energy circuit open; others may or may not.
3. That man-made changes are of a different order than evolutionary changes, and have effects more comprehensive than is intended or foreseen.

These ideas, collectively, raise two basic issues: Can the land adjust itself to the new order? Can the desired alterations be accomplished with less violence?

Biotas seem to differ in their capacity to sustain violent conversion. Western Europe, for example, carries a far different pyramid than Caesar found there. Some large animals are lost; swampy forests have become meadows or plowland; many plants and animals are introduced, some of which escape as pests; the remaining natives are greatly changed in distribution and abundance. Yet the soil is still there and, with the help of imported nutrients, still fertile; the waters flow normally; the new structure seems to function and to persist. There is no visible stoppage or derangement of the circuit.

Western Europe, then, has a resistant biota. Its inner processes are tough, elastic, resistant to strain. No matter how violent the alterations, the pyramid, so far, has developed some new *modus vivendi* which preserves its habitability for man, and for most of the other natives.

Japan seems to present another instance of radical conversion without disorganization.

Most other civilized regions, and some as yet barely touched by civilization, display various stages of disorganization, varying from initial symptoms to advanced wastage. In Asia Minor and North Africa diagnosis is confused by climatic changes, which may have been either the cause or the effect of advanced wastage. In the United States the degree of disorganization varies locally; it is worst in the Southwest, the Ozarks, and parts of the South, and least in New England and the Northwest. Better land-uses may still arrest it in the less advanced regions. In parts of Mexico, South America, South Africa, and Australia a violent and accelerating wastage is in progress, but I cannot assess the prospects.

This almost world-wide display of disorganization in the land seems to be similar to disease in an animal, except that it never culminates in complete disorganization or death. The land recovers, but at some reduced level of complexity, and with a reduced carrying capacity for people, plants, and animals. Many biotas currently regarded as "lands of opportunity" are in fact already subsisting on exploitative agriculture, i.e., they have already exceeded their sustained carrying capacity. Most of South America is overpopulated in this sense.

In arid regions we attempt to offset the process of wastage by reclamation, but it is only too evident that the prospective longevity of reclamation projects is often short. In our own West, the best of them may not last a century.

The combined evidence of history and ecology seems to support one general deduction: the less violent the man-made changes, the greater the probability of successful readjustment in the pyramid. Violence, in turn, varies with human population density; a dense population requires a more violent conversion. In this respect, North America has a better chance for permanence than Europe, if she can contrive to limit her density.

This deduction runs counter to our current philosophy, which assumes that because a small increase in density enriched human life, that an indefinite increase will enrich it indefinitely. Ecology knows of no density relationship that holds for indefinitely wide limits. All gains from density are subject to a law of diminishing returns.

Whatever may be the equation for men and land, it is improbable that we as yet know all its terms. Recent discoveries in mineral and vitamin nutrition reveal unsuspected dependencies in the up-circuit: incredibly minute quantities of certain substances determine the value of soils to plants, of plants to animals. What of the down-circuit? What of the vanishing species, the preservation of which we now regard as an esthetic luxury? They helped build the soil; in what unsuspected ways may they be essential to its maintenance? Professor Weaver proposes that we use prairie flowers to reflocculate the wasting soils of the dust bowl; who knows for what purpose cranes and condors, otters and grizzlies may some day be used?

Land Health and the A-B Cleavage

A land ethic, then, reflects the existence of an ecological conscience, and this in turn reflects a conviction of individual responsibility for the health of the land. Health is the capacity of the land for self-renewal. Conservation is our effort to understand and preserve this capacity.

Conservationists are notorious for their dissensions. Superficially these seem to add up to mere confusion, but a more careful scrutiny reveals a single plane of cleavage common to many specialized fields. In each field one group (A) regards the land as soil, and its function as commodity-production; another group (B) regards the land as a biota, and its function as something broader. How much broader is admittedly in a state of doubt and confusion.

In my own field, forestry, group A is quite content to grow trees like cabbages, with cellulose as the basic forest commodity. It feels no inhibition against violence; its ideology is agronomic. Group B, on the other hand, sees forestry as fundamentally different from agronomy because it employs natural species, and manages a natural environment rather than creating an artificial one. Group B prefers natural reproduction on principle. It worries on biotic as well as economic grounds about the loss of species like chestnut, and the threatened loss of the white pines. It worries about a whole series of secondary forest functions: wildlife, recreation, watersheds, wilderness areas. To my mind, Group B feels the stirrings of an ecological conscience.

In the wildlife field, a parallel cleavage exists. For Group A the basic commodities are sport and meat; the yardsticks of production are ciphers of take in pheasants and trout. Artificial propagation is acceptable as a permanent as well as a temporary recourse—if its unit costs permit. Group B, on the other hand, worries about a whole series of biotic side-issues. What is the cost in predators of producing a game crop? Should we have further recourse to exotics? How can management restore the shrinking species, like prairie grouse, already hopeless as shootable game? How can management restore the threatened rarities, like trumpeter swan and whooping crane? Can management principles be extended to wildflowers? Here again it is clear to me that we have the same A-B cleavage as in forestry.

In the larger field of agriculture I am less competent to speak, but there seem to be somewhat parallel cleavages. Scientific agriculture was actively developing before ecology was born, hence a slower penetration of ecological concepts might be expected. Moreover the farmer, by the very nature of his techniques, must modify the biota more radically than the forester or the wildlife manager. Nevertheless, there are many discontents in agriculture which seem to add up to a new vision of "biotic farming."

Perhaps the most important of these is the new evidence that poundage or tonnage is no measure of the food-value of farm crops;

the products of fertile soil may be qualitatively as well as quantitatively superior. We can bolster poundage from depleted soils by pouring on imported fertility, but we are not necessarily bolstering food-value. The possible ultimate ramifications of this idea are so immense that I must leave their exposition to abler pens.

The discontent that labels itself "organic farming," while bearing some of the earmarks of a cult, is nevertheless biotic in its direction, particularly in its insistence on the importance of soil flora and fauna.

The ecological fundamentals of agriculture are just as poorly known to the public as in other fields of land-use. For example, few educated people realize that the marvelous advances in technique made during recent decades are improvements in the pump, rather than the well. Acre for acre, they have barely sufficed to offset the sinking level of fertility.

In all of these cleavages, we see repeated the same basic paradoxes: man the conqueror *versus* man the biotic citizen; science the sharpener of his sword *versus* science the search-light of his universe; land the slave and servant *versus* land the collective organism. Robinson's injunction to Tristram may well be applied, at this juncture, to *Homo sapiens* as a species in geological time:

> Whether you will or not
> You are a King, Tristram, for you are one
> Of the time-tested few that leave the world,
> When they are gone, not the same place it was.
> Mark what you leave.

The Outlook

It is inconceivable to me that an ethical relation to land can exist without love, respect, and admiration for land, and a high regard for its value. By value, I of course mean something far broader than mere economic value; I mean value in the philosophical sense.

Perhaps the most serious obstacle impeding the evolution of a land ethic is the fact that our educational and economic system is headed away from, rather than toward, an intense consciousness of land. Your true modern is separated from the land by many middlemen, and by innumerable physical gadgets. He has no vital relation to it; to him it is the space between cities on which crops grow. Turn him loose for a day on the land, and if the spot does not happen to be a golf links or a "scenic" area, he is bored stiff. If crops could be raised by hydroponics instead of farming, it would suit him very well. Synthetic substitutes for wood, leather, wool, and other natural land products suit him better than the originals. In short, land is something he has "outgrown."

Almost equally serious as an obstacle to a land ethic is the attitude of the farmer for whom the land is still an adversary, or a taskmaster that keeps him in slavery. Theoretically, the mechanization of farming ought to cut the farmer's chains, but whether it really does is debatable.

One of the requisites for an ecological comprehension of land is an understanding of ecology, and this is by no means co-extensive with "education"; in fact, much higher education seems deliberately to avoid ecological concepts. An understanding of ecology does not necessarily originate in courses bearing ecological labels; it is quite as likely to be labeled geography, botany, agronomy, history, or economics. This is as it should be, but whatever the label, ecological training is scarce.

The case for a land ethic would appear hopeless but for the minority which is in obvious revolt against these "modern" trends.

The "key-log" which must be moved to release the evolutionary process for an ethic is simply this: quit thinking about decent land-use as solely an economic problem. Examine each question in terms of what is ethically and esthetically right, as well as what is economically expedient. A thing is right when it tends to preserve the integrity, stability, and beauty of the biotic community. It is wrong when it tends otherwise.

It of course goes without saying that economic feasibility limits the tether of what can or cannot be done for land. It always has and it always will. The fallacy the economic determinists have tied around our collective neck, and which we now need to cast off, is the belief that economics determines *all* land-use. This is simply not true. An innumerable host of actions and attitudes, comprising perhaps the bulk of all land relations, is determined by the land-users' tastes and predilections, rather than by his purse. The bulk of all land relations hinges on investments of time, forethought, skill, and faith rather than on investments of cash. As a land-user thinketh, so is he.

I have purposely presented the land ethic as a product of social evolution because nothing so important as an ethic is ever "written." Only the most superficial student of history supposes that Moses "wrote" the Decalogue; it evolved in the minds of a thinking community, and Moses wrote a tentative summary of it for a "seminar." I say tentative because evolution never stops.

The evolution of a land ethic is an intellectual as well as emotional process. Conservation is paved with good intentions which prove to be futile, or even dangerous, because they are devoid of critical understanding either of the land, or of economic land-use. I think it is a truism that as the ethical frontier advances from the individual to the community, its intellectual content increases.

The mechanism of operation is the same for any ethic: social approbation for right actions: social disapproval for wrong actions.

By and large, our present problem is one of attitudes and implements. We are remodeling the Alhambra with a steamshovel, and we are proud of our yardage. We shall hardly relinquish the shovel, which after all has many good points, but we are in need of gentler and more objective criteria for its successful use.

II THE CONTEMPORARY CONTROVERSY BEGINS

Rachel Carson

BIOLOGY OR OBLIVION?

When Silent Spring *appeared in 1962, Rachel Carson (1907–1964) was already famous for her eloquent and persuasive naturalistic studies,* Edge of the Sea *(1955),* The Sea Around Us *(1951), and* Under the Sea Wind *(1941). Trained in genetics and marine biology, and a staff member of the Marine Biological Laboratory at Woods Hole, Massachusetts, Dr. Carson first became troubled about DDT when it was indiscriminately used as a universal pesticide during World War II. By the 1950's, DDT was discovered in fatty tissue of birds and fish even in polar regions, far from any spraying. Dr. Carson became aware of widespread deaths of wild birds from DDT, even while the common household fly had developed immunity to it. She was persuaded to bring the question of pesticides before the general public by E. B. White and William Shawn of* The New Yorker *magazine in 1958, in which a shortened form of her book appeared in the summer of 1960. Dr. Carson publicized the classic DDT chain of concentration, in which elm trees were sprayed, the insecticide concentrated in earthworms, and dosage was lethal to birds who ate the worms. She defined ecology as "the web of life—or death." In the chain or web, injury to one link affected all others. Man's own health and survival would be decisively affected, she argued, by widespread and irresponsible use of pesticides originally intended to kill lower forms of life. Dr. Carson called for the development of natural pest control.*

A Fable for Tomorrow

There was once a town in the heart of America where all life seemed to live in harmony with its surroundings. The town lay in the midst of a checkerboard of prosperous farms, with fields of grain and hillsides of orchards where, in spring, white clouds of bloom drifted above the green fields. In autumn, oak and maple and birch set up a blaze of color that flamed and flickered across a backdrop of pines. Then foxes barked in the hills and deer silently crossed the fields, half hidden in the mists of the fall mornings.

Along the roads, laurel, viburnum and alder, great ferns and wildflowers delighted the traveler's eye through much of the year. Even in winter the roadsides were places of beauty, where countless birds came to feed on the berries and on the seed heads of the dried weeds rising above the snow. The countryside was, in fact, famous

From *Silent Spring* by Rachel Carson. Copyright © 1962 by Rachel L. Carson. Reprinted by permission of the publisher, Houghton Mifflin Company, and Marie Rodell.

for the abundance and variety of its bird life, and when the flood of migrants was pouring through in spring and fall people traveled from great distances to observe them. Others came to fish the streams, which flowed clear and cold out of the hills and contained shady pools where trout lay. So it had been from the days many years ago when the first settlers raised their houses, sank their wells, and built their barns.

Then a strange blight crept over the area and everything began to change. Some evil spell had settled on the community; mysterious maladies swept the flocks of chickens; the cattle and sheep sickened and died. Everywhere was a shadow of death. The farmers spoke of much illness among their families. In the town the doctors had become more and more puzzled by new kinds of sickness appearing among their patients. There had been several sudden and unexplained deaths, not only among adults but even among children, who would be stricken suddenly while at play and die within a few hours.

There was a strange stillness. The birds, for example—where had they gone? Many people spoke of them, puzzled and disturbed. The feeding stations in the backyards were deserted. The few birds seen anywhere were moribund; they trembled violently and could not fly. It was a spring without voices. On the mornings that had once throbbed with the dawn chorus of robins, catbirds, doves, jays, wrens, and scores of other bird voices there was now no sound; only silence lay over the fields and woods and marsh.

On the farms the hens brooded, but no chicks hatched. The farmers complained that they were unable to raise any pigs—the litters were small and the young survived only a few days. The apple trees were coming into bloom but no bees droned among the blossoms, so there was no pollination and there would be no fruit.

The roadsides, once so attractive, were now lined with browned and withered vegetation as though swept by fire. These, too, were silent, deserted by all living things. Even the streams were now lifeless. Anglers no longer visited them, for all the fish had died.

In the gutters under the eaves and between the shingles of the roofs, a white granular powder still showed a few patches; some weeks before it had fallen like snow upon the roofs and the lawns, the fields and streams.

No witchcraft, no enemy action had silenced the rebirth of new life in this stricken world. The people had done it themselves.

This town does not actually exist, but it might easily have a thousand counterparts in America or elsewhere in the world. I know of no community that has experienced all the misfortunes I describe. Yet every one of these disasters has actually happened somewhere, and many real communities have already suffered a substantial number of them. A grim specter has crept upon us almost unnoticed, and this imagined tragedy may easily become a stark reality we all shall know.

What has already silenced the voices of spring in countless towns in America? This book is an attempt to explain.

The Obligation to Endure

The history of life on earth has been a history of interaction between living things and their surroundings. To a large extent, the physical form and the habits of the earth's vegetation and its animal life have been molded by the environment. Considering the whole span of earthly time, the opposite effect, in which life actually modifies its surroundings, has been relatively slight. Only within the moment of time represented by the present century has one species—man— acquired significant power to alter the nature of his world.

During the past quarter century this power has not only increased to one of disturbing magnitude but it has changed in character. The most alarming of all man's assaults upon the environment is the contamination of air, earth, rivers, and sea with dangerous and even lethal materials. This pollution is for the most part irrecoverable; the chain of evil it initiates not only in the world that must support life but in living tissues is for the most part irreversible. In this now universal contamination of the environment, chemicals are the sinister and little-recognized partners of radiation in changing the very nature of the world—the very nature of its life. Strontium 90, released through nuclear explosions into the air, comes to earth in rain or drifts down as fallout, lodges in soil, enters into the grass or corn or wheat grown there, and in time takes up its abode in the bones of a human being, there to remain until his death. Similarly, chem-

icals sprayed on croplands or forests or gardens lie long in soil,
entering into living organisms, passing from one to another in a
chain of poisoning and death. Or they pass mysteriously by under-
ground streams until they emerge and, through the alchemy of air
and sunlight, combine into new forms that kill vegetation, sicken
cattle, and work unknown harm on those who drink from once-pure
wells. As Albert Schweitzer has said, "Man can hardly even recognize the devils of his own creation."

It took hundreds of millions of years to produce the life that now
inhabits the earth—eons of time in which that developing and evolv-
ing and diversifying life reached a state of adjustment and balance
with its surroundings. The environment, rigorously shaping and di-
recting the life it supported, contained elements that were hostile as
well as supporting. Certain rocks gave out dangerous radiation; even
within the light of the sun, from which all life draws its energy, there
were short-wave radiations with power to injure. Given time—time
not in years but in millennia—life adjusts, and a balance has been
reached. For time is the essential ingredient; but in the modern world
there is no time.

The rapidity of change and the speed with which new situations
are created follow the impetuous and heedless pace of man rather
than the deliberate pace of nature. Radiation is no longer merely the
background radiation of rocks, the bombardment of cosmic rays, the
ultraviolet of the sun that have existed before there was any life on
earth; radiation is now the unnatural creation of man's tampering
with the atom. The chemicals to which life is asked to make its adjustment are no longer merely the calcium and silica and copper and
all the rest of the minerals washed out of the rocks and carried in
rivers to the sea; they are the synthetic creations of man's inventive
mind, brewed in his laboratories, and having no counterparts in
nature.

To adjust to these chemicals would require time on the scale that
is nature's; it would require not merely the years of a man's life but
the life of generations. And even this, were it by some miracle possible, would be futile, for the new chemicals come from our laboratories in an endless stream; almost five hundred annually find their
way into actual use in the United States alone. The figure is staggering and its implications are not easily grasped—500 new chemicals

to which the bodies of men and animals are required somehow to adapt each year, chemicals totally outside the limits of biologic experience.

Among them are many that are used in man's war against nature. Since the mid-1940's over 200 basic chemicals have been created for use in killing insects, weeds, rodents, and other organisms described in the modern vernacular as "pests"; and they are sold under several thousand different brand names.

These sprays, dusts, and aerosols are now applied almost universally to farms, gardens, forests, and homes—nonselective chemicals that have the power to kill every insect, the "good" and the "bad," to still the song of birds and the leaping of fish in the streams, to coat the leaves with a deadly film, and to linger on in soil—all this though the intended target may be only a few weeds or insects. Can anyone believe it is possible to lay down such a barrage of poisons on the surface of the earth without making it unfit for all life? They should not be called "insecticides," but "biocides."

The whole process of spraying seems caught up in an endless spiral. Since DDT was released for civilian use, a process of escalation has been going on in which ever more toxic materials must be found. This has happened because insects, in a triumphant vindication of Darwin's principle of the survival of the fittest, have evolved super races immune to the particular insecticide used, hence a deadlier one has always to be developed—and then a deadlier one than that. It has happened also because, for reasons to be described later, destructive insects often undergo a "flareback," or resurgence, after spraying, in numbers greater than before. Thus the chemical war is never won, and all life is caught in its violent crossfire.

Along with the possibility of the extinction of mankind by nuclear war, the central problem of our age has therefore become the contamination of man's total environment with such substances of incredible potential for harm—substances that accumulate in the tissues of plants and animals and even penetrate the germ cells to shatter or alter the very material of heredity upon which the shape of the future depends.

Some would-be architects of our future look toward a time when it will be possible to alter the human germ plasm by design. But we may easily be doing so now by inadvertence, for many chemicals, like radiation, bring about gene mutations. It is ironic to think that man might determine his own future by something so seemingly trivial as the choice of an insect spray.

All this has been risked—for what? Future historians may well be amazed by our distorted sense of proportion. How could intelligent beings seek to control a few unwanted species by a method that contaminated the entire environment and brought the threat of disease and death even to their own kind? Yet this is precisely what we have done. We have done it, moreover, for reasons that collapse the moment we examine them. We are told that the enormous and expanding use of pesticides is necessary to maintain farm production. Yet is our real problem not one of *overproduction?* Our farms, despite measures to remove acreages from production and to pay farmers *not* to produce, have yielded such a staggering excess of crops that the American taxpayer in 1962 is paying out more than one billion dollars a year as the total carrying cost of the surplus-food storage program. And is the situation helped when one branch of the Agriculture Department tries to reduce production while another states, as it did in 1958, "It is believed generally that reduction of crop acreages under provisions of the Soil Bank will stimulate interest in use of chemicals to obtain maximum production on the land retained in crops."

All this is not to say there is no insect problem and no need of control. I am saying, rather, that control must be geared to realities, not to mythical situations, and that the methods employed must be such that they do not destroy us along with the insects.

The problem whose attempted solution has brought such a train of disaster in its wake is an accompaniment of our modern way of life. Long before the age of man, insects inhabited the earth—a group of extraordinarily varied and adaptable beings. Over the course of time since man's advent, a small percentage of the more than half a million species of insects have come into conflict with human welfare in two principal ways: as competitors for the food supply and as carriers of human disease.

Disease-carrying insects become important where human beings are crowded together, especially under conditions where sanitation is poor, as in time of natural disaster or war or in situations of extreme poverty and deprivation. Then control of some sort becomes necessary. It is a sobering fact, however, as we shall presently see, that the method of massive chemical control has had only limited success, and also threatens to worsen the very conditions it is intended to curb.

Under primitive agricultural conditions the farmer had few insect problems. These arose with the intensification of agriculture—the devotion of immense acreages to a single crop. Such a system set the stage for explosive increases in specific insect populations. Single-crop farming does not take advantage of the principles by which nature works; it is agriculture as an engineer might conceive it to be. Nature has introduced great variety into the landscape, but man has displayed a passion for simplifying it. Thus he undoes the built-in checks and balances by which nature holds the species within bounds. One important natural check is a limit on the amount of suitable habitat for each species. Obviously then, an insect that lives on wheat can build up its population to much higher levels on a farm devoted to wheat than on one in which wheat is intermingled with other crops to which the insect is not adapted.

The same thing happens in other situations. A generation or more ago, the towns of large areas of the United States lined their streets with the noble elm tree. Now the beauty they hopefully created is threatened with complete destruction as disease sweeps through the elms, carried by a beetle that would have only limited chance to build up large populations and to spread from tree to tree if the elms were only occasional trees in a richly diversified planting.

Another factor in the modern insect problem is one that must be viewed against a background of geologic and human history: the spreading of thousands of different kinds of organisms from their native homes to invade new territories. This worldwide migration has been studied and graphically described by the British ecologist Charles Elton in his recent book *The Ecology of Invasions.* During the Cretaceous Period, some hundred million years ago, flooding seas cut many land bridges between continents and living things found themselves confined in what Elton calls "colossal separate nature reserves." There, isolated from others of their kind, they developed many new species. When some of the land masses were joined again, about 15 million years ago, these species began to move out into new territories—a movement that is not only still in progress but is now receiving considerable assistance from man.

The importation of plants is the primary agent in the modern spread of species, for animals have almost invariably gone along with the plants, quarantine being a comparatively recent and not completely effective innovation. The United States Office of Plant Introduction alone has introduced almost 200,000 species and varieties of plants from all over the world. Nearly half of the 180 or so major insect enemies of plants in the United States are accidental imports from abroad, and most of them have come as hitchhikers on plants.

In new territory, out of reach of the restraining hand of the natural enemies that kept down its numbers in its native land, an invading plant or animal is able to become enormously abundant. Thus it is no accident that our most troublesome insects are introduced species.

These invasions, both the naturally occurring and those dependent on human assistance, are likely to continue indefinitely. Quarantine and massive chemical campaigns are only extremely expensive ways of buying time. We are faced, according to Dr. Elton, "with a life-and-death need not just to find new technological means of suppressing this plant or that animal"; instead we need the basic knowledge of animal populations and their relations to their surroundings that will "promote an even balance and damp down the explosive power of outbreaks and new invasions."

Much of the necessary knowledge is now available but we do not use it. We train ecologists in our universities and even employ them in our governmental agencies but we seldom take their advice. We allow the chemical death rain to fall as though there were no alternative, whereas in fact there are many, and our ingenuity could soon discover many more if given opportunity.

Have we fallen into a mesmerized state that makes us accept as inevitable that which is inferior or detrimental, as though having lost the will or the vision to demand that which is good? Such thinking, in the words of the ecologist Paul Shepard, "idealizes life with only its head out of water, inches above the limits of toleration of the corruption of its own environment. . . . Why should we tolerate a diet of weak poisons, a home in insipid surroundings, a circle of acquaintances who are not quite our enemies, the noise of motors with just enough relief to prevent insanity? Who would want to live in a world which is just not quite fatal?"

Yet such a world is pressed upon us. The crusade to create a chemically sterile, insect-free world seems to have engendered a fanatic zeal on the part of many specialists and most of the so-called control agencies. On every hand there is evidence that those engaged in spraying operations exercise a ruthless power. "The regulatory entomologists . . . function as prosecutor, judge and jury, tax assessor and collector and sheriff to enforce their own orders," said Connecticut entomologist Neely Turner. The most flagrant abuses go unchecked in both state and federal agencies.

It is not my contention that chemical insecticides must never be used. I do contend that we have put poisonous and biologically potent chemicals indiscriminately into the hands of persons largely or wholly ignorant of their potentials for harm. We have subjected enormous numbers of people to contact with these poisons, without their consent and often without their knowledge. If the Bill of Rights contains no guarantee that a citizen shall be secure against lethal poisons distributed either by private individuals or by public officials, it is surely only because our forefathers, despite their considerable wisdom and foresight, could conceive of no such problem.

I contend, furthermore, that we have allowed these chemicals to be used with little or no advance investigation of their effect on soil, water, wildlife, and man himself. Future generations are unlikely to condone our lack of prudent concern for the integrity of the natural world that supports all life.

There is still very limited awareness of the nature of the threat. This is an era of specialists, each of whom sees his own problem and is unaware of or intolerant of the larger frame into which it fits. It is also an era dominated by industry, in which the right to make a dollar at whatever cost is seldom challenged. When the public protests, confronted with some obvious evidence of damaging results of pesticide applications, it is fed little tranquilizing pills of half truth. We urgently need an end to these false assurances, to the sugar coating of unpalatable facts. It is the public that is being asked to assume the risks that the insect controllers calculate. The public must decide whether it wishes to continue on the present road, and it can do so only when in full possession of the facts. In the words of Jean Rostand, "The obligation to endure gives us the right to know."

*   *   *

Needless Havoc

As man proceeds toward his announced goal of the conquest of nature, he has written a depressing record of destruction, directed not only against the earth he inhabits but against the life that shares it with him. The history of the recent centuries has its black passages —the slaughter of the buffalo on the western plains, the massacre of the shorebirds by the market gunners, the near-extermination of the egrets for their plumage. Now, to these and others like them, we are adding a new chapter and a new kind of havoc—the direct killing of birds, mammals, fishes, and indeed practically every form of wildlife by chemical insecticides indiscriminately sprayed on the land.

Under the philosophy that now seems to guide our destinies, nothing must get in the way of the man with the spray gun. The incidental victims of his crusade against insects count as nothing; if robins, pheasants, raccoons, cats, or even livestock happen to inhabit the same bit of earth as the target insects and to be hit by the rain of insect-killing poisons no one must protest.

The citizen who wishes to make a fair judgment of the question of wildlife loss is today confronted with a dilemma. On the one hand conservationists and many wildlife biologists assert that the losses have been severe and in some cases even catastrophic. On the other hand the control agencies tend to deny flatly and categorically that such losses have occurred, or that they are of any importance if they have. Which view are we to accept?

The credibility of the witness is of first importance. The professional wildlife biologist on the scene is certainly best qualified to discover and interpret wildlife loss. The entomologist, whose specialty is insects, is not so qualified by training, and is not psychologically disposed to look for undesirable side effects of his control program. Yet it is the control men in state and federal governments —and of course the chemical manufacturers—who steadfastly deny the facts reported by the biologists and declare they see little evidence of harm to wildlife. Like the priest and the Levite in the biblical story, they choose to pass by on the other side and to see nothing. Even if we charitably explain their denials as due to the shortsightedness of the specialist and the man with an interest this does not mean we must accept them as qualified witnesses.

The best way to form our own judgment is to look at some of the major control programs and learn, from observers familiar with the ways of wildlife, and unbiased in favor of chemicals, just what has happened in the wake of a rain of poison falling from the skies into the world of wildlife.

To the bird watcher, the suburbanite who derives joy from birds in his garden, the hunter, the fisherman or the explorer of wild regions, anything that destroys the wildlife of an area for even a single year has deprived him of pleasure to which he has a legitimate right. This is a valid point of view. Even if, as has sometimes happened, some of the birds and mammals and fishes are able to re-establish themselves after a single spraying, a great and real harm has been done.

But such re-establishment is unlikely to happen. Spraying tends to be repetitive, and a single exposure from which the wildlife populations might have a chance to recover is a rarity. What usually results is a poisoned environment, a lethal trap in which not only the resident populations succumb but those who come in as migrants as well. The larger the area sprayed the more serious the harm, because no oases of safety remain. Now, in a decade marked by insect-control programs in which many thousands or even millions of acres are sprayed as a unit, a decade in which private and community spraying has also surged steadily upward, a record of destruction and death of American wildlife has accumulated. Let us look at some of these programs and see what has happened.

During the fall of 1959 some 27,000 acres in southeastern Michigan, including numerous suburbs of Detroit, were heavily dusted from the air with pellets of aldrin, one of the most dangerous of all the chlorinated hydrocarbons. The program was conducted by the Michigan Department of Agriculture with the cooperation of the United States Department of Agriculture; its announced purpose was control of the Japanese beetle.

Little need was shown for this drastic and dangerous action. On the contrary, Walter P. Nickell, one of the best-known and best-informed naturalists in the state, who spends much of his time in the field with long periods in southern Michigan every summer, declared: "For more than thirty years, to my direct knowledge, the Japanese beetle has been present in the city of Detroit in small numbers. The numbers have not shown any appreciable increase in all this lapse of years. I have yet to see a single Japanese beetle [in 1959] other than the few caught in Government catch traps in Detroit. . . . Everything is being kept so secret that I have not yet been able to obtain any information whatsoever to the effect that they have increased in numbers."

An official release by the state agency merely declared that the beetle had "put in its appearance" in the areas designated for the aerial attack upon it. Despite the lack of justification the program was launched, with the state providing the manpower and supervising the operation, the federal government providing equipment and additional men, and the communities paying for the insecticide.

The Japanese beetle, an insect accidentally imported into the United States, was discovered in New Jersey in 1916, when a few shiny beetles of a metallic green color were seen in a nursery near Riverton. The beetles, at first unrecognized, were finally identified as a common inhabitant of the main islands of Japan. Apparently they had entered the United States on nursery stock imported before restrictions were established in 1912.

From its original point of entrance the Japanese beetle has spread rather widely throughout many of the states east of the Mississippi, where conditions of temperature and rainfall are suitable for it. Each year some outward movement beyond the existing boundaries of its distribution usually takes place. In the eastern areas where the beetles have been longest established, attempts have been made to set up natural controls. Where this has been done, the beetle populations have been kept at relatively low levels, as many records attest.

Despite the record of reasonable control in eastern areas, the midwestern states now on the fringe of the beetle's range have launched an attack worthy of the most deadly enemy instead of only a moderately destructive insect, employing the most dangerous chemicals distributed in a manner that exposes large numbers of people, their domestic animals, and all wildlife to the poison intended for the beetle. As a result these Japanese beetle programs have caused shocking destruction of animal life and have exposed human beings to undeniable hazard. Sections of Michigan, Kentucky, Iowa, Indiana, Illinois, and Missouri are all experiencing a rain of chemicals in the name of beetle control.

The Michigan spraying was one of the first large-scale attacks on the Japanese beetle from the air. The choice of aldrin, one of the deadliest of all chemicals, was not determined by any peculiar suitability for Japanese beetle control, but simply by the wish to save money—aldrin was the cheapest of the compounds available. While the state in its official release to the press acknowledged that aldrin is a "poison," it implied that no harm could come to human beings in the heavily populated areas to which the chemical was applied. (The official answer to the query "What precautions should I take?" was "For you, none.") An official of the Federal Aviation Agency was later quoted in the local press to the effect that "this is a safe operation" and a representative of the Detroit Department of Parks and Recreation added his assurance that "the dust is harmless to humans and will not hurt plants or pets." One must assume that none of these officials had consulted the published and readily available reports of the United States Public Health Service, the Fish and Wildlife Service, and other evidence of the extremely poisonous nature of aldrin.

Acting under the Michigan pest control law which allows the state to spray indiscriminately without notifying or gaining permission of individual landowners, the low-flying planes began to fly over the Detroit area. The city authorities and the Federal Aviation Agency were immediately besieged by calls from worried citizens. After receiving nearly 800 calls in a single hour, the police begged radio and television stations and newspapers to "tell the watchers what they were seeing and advise them it was safe," according to the Detroit *News*. The Federal Aviation Agency's safety officer assured the public that "the planes are carefully supervised" and "are authorized to fly low." In a somewhat mistaken attempt to allay fears, he added that the planes had emergency valves that would allow them to dump their entire load instantaneously. This, fortunately, was not done, but as the planes went about their work the pellets of insecticide fell on beetles and humans alike, showers of "harmless" poison descending on people shopping or going to work and on children out from school for the lunch hour. Housewives swept the granules from porches and sidewalks, where they are said to have "looked like snow." As pointed out later by the Michigan Audubon Society, "In the spaces between shingles on roofs, in eaves-troughs, in the cracks in bark and twigs, the little white pellets of aldrin-and-clay, no bigger than a pin head, were lodged by the millions. . . . When the snow and rain came, every puddle became a possible death potion."

Within a few days after the dusting operation, the Detroit Audubon Society began receiving calls about the birds. According to the Society's secretary, Mrs. Ann Boyes, "The first indication that the people were concerned about the spray was a call I received on Sunday morning from a woman who reported that coming home from church she saw an alarming number of dead and dying birds. The spraying there had been done on Thursday. She said there were no birds at all flying in the area, that she had found at least a dozen [dead] in her backyard and that the neighbors had found dead squirrels." All other calls received by Mrs. Boyes that day reported "a great many dead birds and no live ones. . . . People who had maintained bird feeders said there were no birds at all at their feeders." Birds picked up in a dying condition showed the typical symptoms of insecticide poisoning—tremoring, loss of ability to fly, paralysis, convulsions.

Nor were birds the only forms of life immediately affected. A local veterinarian reported that his office was full of clients with dogs and cats that had suddenly sickened. Cats, who so meticulously groom their coats and lick their paws, seemed to be most affected. Their illness took the form of severe diarrhea, vomiting and convulsions. The only advice the veterinarian could give his clients was not to let the animals out unnecessarily, or to wash the paws promptly if they did so. (But the chlorinated hydrocarbons cannot be washed even from fruits or vegetables, so little protection could be expected from this measure.)

Despite the insistence of the City–County Health Commissioner that the birds must have been killed by "some other kind of spraying" and that the outbreak of throat and chest irritations that followed the exposure to aldrin must have been due to "something else," the local Health Department received a constant stream of complaints. A prominent Detroit internist was called upon to treat four of his patients within an hour after they had been exposed while watching the planes at work. All had similar symptoms: nausea, vomiting, chills, fever, extreme fatigue, and coughing.

The Detroit experience has been repeated in many other communities as pressure has mounted to combat the Japanese beetle with chemicals. At Blue Island, Illinois, hundreds of dead and dying birds were picked up. Data collected by birdbanders here suggests that 80 percent of the songbirds were sacrificed. In Joliet, Illinois, some 3000 acres were treated with heptachlor in 1959. According to reports from a local sportsmen's club, the bird population within the treated area was "virtually wiped out." Dead rabbits, muskrats, opossums, and fish were also found in numbers, and one of the local schools made the collection of insecticide-poisoned birds a science project.

Perhaps no community has suffered more for the sake of a beetleless world than Sheldon, in eastern Illinois, and adjacent areas in Iroquois County. In 1954 the United States Department of Agriculture and the Illinois Agriculture Department began a program to eradicate the Japanese beetle along the line of its advance into Illinois, holding out the hope, and indeed the assurance, that intensive spraying would destroy the populations of the invading insect. The first "eradication" took place that year, when dieldrin was applied to 1400 acres by air. Another 2600 acres were treated similarly in 1955, and the task was presumably considered complete. But more and more chemical treatments were called for, and by the end of 1961 some 131,000 acres had been covered. Even in the first years of the program it was apparent that heavy losses were occurring among wildlife and domestic animals. The chemical treatments were continued, nevertheless, without consultation with either the United States Fish and Wildlife Service or the Illinois Game Management Division. (In the spring of 1960, however, officials of the federal Department of Agriculture appeared before a congressional committee in opposition to a bill that would require just such prior consultation. They declared blandly that the bill was unnecessary because cooperation and consultation were "usual." These officials were quite unable to recall situations where cooperation had not taken place "at the Washington level." In the same hearings they stated clearly their unwillingness to consult with state fish and game departments.)

Although funds for chemical control came in never-ending streams, the biologists of the Illinois Natural History Survey who attempted to measure the damage to wildlife had to operate on a financial shoestring. A mere $1100 was available for the employment of a field assistant in 1954 and no special funds were provided in 1955. Despite these crippling difficulties, the biologists assembled facts that collectively paint a picture of almost unparalleled wildlife destruction—destruction that became obvious as soon as the program got under way.

Conditions were made to order for poisoning insect-eating birds, both in the poisons used and in the events set in motion by their application. In the early programs at Sheldon, dieldrin was applied at the rate of 3 pounds to the acre. To understand its effect on birds one need only remember that in laboratory experiments on quail dieldrin has proved to be about 50 times as poisonous as DDT. The poison spread over the landscape at Sheldon was therefore roughly equivalent to 150 pounds of DDT per acre! And this was a minimum, because there seems to have been some overlapping of treatments along field borders and in corners.

As the chemical penetrated the soil the poisoned beetle grubs crawled out on the surface of the ground, where they remained for some time before they died, attractive to insect-eating birds. Dead and dying insects of various species were conspicuous for about two weeks after the treatment. The effect on the bird populations could easily have been foretold. Brown thrashers, starlings, meadowlarks, grackles, and pheasants were virtually wiped out. Robins were "almost annihilated," according to the biologists' report. Dead earthworms had been seen in numbers after a gentle rain; probably the robins had fed on the poisoned worms. For other birds, too, the once beneficial rain had been changed, through the evil power of the poison introduced into their world, into an agent of destruction. Birds seen drinking and bathing in puddles left by rain a few days after the spraying were inevitably doomed.

The birds that survived may have been rendered sterile. Although a few nests were found in the treated area, a few with eggs, none contained young birds.

Among the mammals ground squirrels were virtually annihilated; their bodies were found in attitudes characteristic of violent death by poisoning. Dead muskrats were found in the treated areas, dead rabbits in the fields. The fox squirrel had been a relatively common animal in the town; after the spraying it was gone.

It was a rare farm in the Sheldon area that was blessed by the presence of a cat after the war on beetles was begun. Ninety percent of all the farm cats fell victims to the dieldrin during the first season of spraying. This might have been predicted because of the black record of these poisons in other places. Cats are extremely sensitive to all insecticides and especially so, it seems, to dieldrin. In western Java in the course of the anti-malarial program carried out by the World Health Organization, many cats are reported to have died. In central Java so many were killed that the price of a cat more than doubled. Similarly, the World Health Organization, spraying in Venezuela, is reported to have reduced cats to the status of a rare animal.

In Sheldon it was not only the wild creatures and the domestic companions that were sacrificed in the campaign against an insect. Observations on several flocks of sheep and a herd of beef cattle are indicative of the poisoning and death that threatened livestock as well. The Natural History Survey report describes one of these episodes as follows:

The sheep . . . were driven into a small, untreated bluegrass pasture across a gravel road from a field which had been treated with dieldrin spray on May 6. Evidently some spray had drifted across the road into the pasture, for the sheep began to show symptoms of intoxication almost at once. . . . They lost interest in food and displayed extreme restlessness, following the pasture fence around and around apparently searching for a way out. . . . [They] refused to be driven, bleated almost continuously, and stood with their heads lowered; they were finally carried from the pasture. . . . They displayed great desire for water. Two of the sheep were found dead in the stream passing through the pasture, and the remaining sheep were repeatedly driven out of the stream, several having to be dragged forcibly from the water. Three of the sheep eventually died; those remaining recovered to all outward appearances.

This, then, was the picture at the end of 1955. Although the chemical war went on in succeeding years, the trickle of research funds dried up completely. Requests for money for wildlife-insecticide research were included in annual budgets submitted to the Illinois legislature by the Natural History Survey, but were invariably among the first items to be eliminated. It was not until 1960 that money was somehow found to pay the expenses of one field assistant—to do work that could easily have occupied the time of four men.

The desolate picture of wildlife loss had changed little when the biologists resumed the studies broken off in 1955. In the meantime, the chemical had been changed to the even more toxic aldrin, *100 to 300 times* as toxic as DDT in tests on quail. By 1960, every species of wild mammal known to inhabit the area had suffered losses. It was even worse with the birds. In the small town of Donovan the robins had been wiped out, as had the grackles, starlings, and brown thrashers. These and many other birds were sharply reduced elsewhere. Pheasant hunters felt the effects of the beetle campaign sharply. The number of broods produced on treated land fell off by some 50 percent, and the number of young in a brood declined. Pheasant hunting, which had been good in these areas in former years, was virtually abandoned as unrewarding.

In spite of the enormous havoc that had been wrought in the name of eradicating the Japanese beetle, the treatment of more than 100,000 acres in Iroquois County over an eight-year period seems to have resulted in only temporary suppression of the insect, which continues its westward movement. The full extent of the toll that has been taken by this largely ineffective program may never be known, for the results measured by the Illinois biologists are a minimum figure. If the research program had been adequately financed to permit full coverage, the destruction revealed would have been even more appalling. But in the eight years of the program, only about $6000 was provided for biological field studies. Meanwhile the federal government had spent about $375,000 for control work and additional thousands had been provided by the state. The amount spent for research was therefore a small fraction of 1 percent of the outlay for the chemical program.

These midwestern programs have been conducted in a spirit of crisis, as though the advance of the beetle presented an extreme peril justifying any means to combat it. This of course is a distortion of the facts, and if the communities that have endured these chemical drenchings had been familiar with the earlier history of the Japanese beetle in the United States they would surely have been less acquiescent.

The eastern states, which had the good fortune to sustain their beetle invasion in the days before the synthetic insecticides had been invented, have not only survived the invasion but have brought the insect under control by means that represented no threat whatever to other forms of life. There has been nothing comparable to the Detroit or Sheldon sprayings in the East. The effective methods there involved the bringing into play of natural forces of control which have the multiple advantages of permanence and environmental safety.

During the first dozen years after its entry into the United States, the beetle increased rapidly, free of the restraints that in its native land hold it in check. But by 1945 it had become a pest of only minor importance throughout much of the territory over which it had spread. Its decline was largely a consequence of the importation of parasitic insects from the Far East and of the establishment of disease organisms fatal to it.

Between 1920 and 1933, as a result of diligent searching throughout the native range of the beetle, some 34 species of predatory or parasitic insects had been imported from the Orient in an effort to establish natural control. Of these, five became well established in the eastern United States. The most effective and widely distributed is a parasitic wasp from Korea and China, *Tiphia vernalis.* The female *Tiphia,* finding a beetle grub in the soil, injects a paralyzing fluid and attaches a single egg to the undersurface of the grub. The young wasp, hatching as a larva, feeds on the paralyzed grub and destroys it. In some 25 years, colonies of *Tiphia* were introduced into 14 eastern states in a cooperative program of state and federal agencies. The wasp became widely established in this area and is generally credited by entomologists with an important role in bringing the beetle under control.

An even more important role has been played by a bacterial disease that affects beetles of the family to which the Japanese beetle belongs—the scarabaeids. It is a highly specific organism, attacking no other type of insects, harmless to earthworms, warm-blooded animals, and plants. The spores of the disease occur in soil. When ingested by a foraging beetle grub they multiply prodigiously in its blood, causing it to turn an abnormally white color, hence the popular name, "milky disease."

Milky disease was discovered in New Jersey in 1933. By 1938 it was rather widely prevalent in the older areas of Japanese beetle infestation. In 1939 a control program was launched, directed at speeding up the spread of the disease. No method had been developed for growing the disease organism in an artificial medium, but a satisfactory substitute was evolved; infected grubs are ground up, dried, and combined with chalk. In the standard mixture a gram of dust contains 100 million spores. Between 1939 and 1953 some 94,000 acres in 14 eastern states were treated in a cooperative federal-state program; other areas on federal lands were treated; and an unknown but extensive area was treated by private organizations or individuals. By 1945, milky spore disease was raging among the beetle populations of Connecticut, New York, New Jersey, Delaware, and Maryland. In some test areas infection of grubs had reached as high as 94 percent. The distribution program was discontinued as a governmental enterprise in 1953 and production was taken over by a private laboratory, which continues to supply individuals, garden clubs, citizens' associations, and all others interested in beetle control.

The eastern areas where this program was carried out now enjoy a high degree of natural protection from the beetle. The organism remains viable in the soil for years and therefore becomes to all intents and purposes permanently established, increasing in effectiveness, and being continuously spread by natural agencies.

Why, then, with this impressive record in the East, were the same procedures not tried in Illinois and the other midwestern states where the chemical battle of the beetles is now being waged with such fury?

We are told that inoculation with milky spore disease is "too expensive"—although no one found it so in the 14 eastern states in the 1940's. And by what sort of accounting was the "too expensive" judgment reached? Certainly not by any that assessed the true costs of the total destruction wrought by such programs as the Sheldon spraying. This judgment also ignores the fact that inoculation with the spores need be done only once; the first cost is the only cost.

We are told also that milky spore disease cannot be used on the periphery of the beetle's range because it can be established only where a large grub population is *already* present in the soil. Like many other statements in support of spraying, this one needs to be questioned. The bacterium that causes milky spore disease has been found to infect at least 40 other species of beetles which collectively have quite a wide distribution and would in all probability serve to establish the disease even where the Japanese beetle population is very small or nonexistent. Furthermore, because of the long viability of the spores in soil they can be introduced even in the complete absence of grubs, as on the fringe of the present beetle infestation, there to await the advancing population.

Those who want immediate results, at whatever cost, will doubtless continue to use chemicals against the beetle. So will those who favor the modern trend to built-in obsolescence, for chemical control is self-perpetuating, needing frequent and costly repetition.

On the other hand, those who are willing to wait an extra season or two for full results will turn to milky disease; they will be rewarded with lasting control that becomes more, rather than less effective with the passage of time.

An extensive program of research is under way in the United States Department of Agriculture laboratory at Peoria, Illinois, to find a way to culture the organism of milky disease on an artificial medium. This will greatly reduce its cost and should encourage its more extensive use. After years of work, some success has now been reported. When this "breakthrough" is thoroughly established perhaps some sanity and perspective will be restored to our dealings with the Japanese beetle, which at the peak of its depredations never justified the nightmare excesses of some of these midwestern programs.

Incidents like the eastern Illinois spraying raise a question that is not only scientific but moral. The question is whether any civilization can wage relentless war on life without destroying itself, and without losing the right to be called civilized.

These insecticides are not selective poisons; they do not single out the one species of which we desire to be rid. Each of them is used for the simple reason that it is a deadly poison. It therefore poisons all life with which it comes in contact: the cat beloved of some family, the farmer's cattle, the rabbit in the field, and the horned lark out of the sky. These creatures are innocent of any harm to man. Indeed, by their very existence they and their fellows make his life more pleasant. Yet he rewards them with a death that is not only sudden but horrible. Scientific observers at Sheldon described the symptoms of a meadowlark found near death: "Although it lacked muscular coordination and could not fly or stand, it continued to beat its wings and clutch with its toes while lying on its side. Its beak was held open and breathing was labored." Even more pitiful was the mute testimony of the dead ground squirrels, which "exhibited a characteristic attitude in death. The back was bowed, and the forelegs with the toes of the feet tightly clenched were drawn close to the thorax. . . . The head and neck were outstretched and the mouth often contained dirt, suggesting that the dying animal had been biting at the ground."

By acquiescing in an act that can cause such suffering to a living creature, who among us is not diminished as a human being?

*   *   *

The Other Road

We stand now where two roads diverge. But unlike the roads in Robert Frost's familiar poem, they are not equally fair. The road we have long been traveling is deceptively easy, a smooth superhighway on which we progress with great speed, but at its end lies disaster. The other fork of the road—the one "less traveled by"—

offers our last, our only chance to reach a destination that assures the preservation of our earth.

The choice, after all, is ours to make. If, having endured much, we have at last asserted our "right to know," and if, knowing, we have concluded that we are being asked to take senseless and frightening risks, then we should no longer accept the counsel of those who tell us that we must fill our world with poisonous chemicals; we should look about and see what other course is open to us.

A truly extraordinary variety of alternatives to the chemical control of insects is available. Some are already in use and have achieved brilliant success. Others are in the stage of laboratory testing. Still others are little more than ideas in the minds of imaginative scientists, waiting for the opportunity to put them to the test. All have this in common: they are *biological* solutions, based on understanding of the living organisms they seek to control, and of the whole fabric of life to which these organisms belong. Specialists representing various areas of the vast field of biology are contributing—entomologists, pathologists, geneticists, physiologists, biochemists, ecologists—all pouring their knowledge and their creative inspirations into the formation of a new science of biotic controls.

"Any science may be likened to a river," says a Johns Hopkins biologist, Professor Carl P. Swanson. "It has its obscure and unpretentious beginning; its quiet stretches as well as its rapids; its periods of drought as well as of fullness. It gathers momentum with the work of many investigators and as it is fed by other streams of thought; it is deepened and broadened by the concepts and generalizations that are gradually evolved."

So it is with the science of biological control in its modern sense. In America it had its obscure beginnings a century ago with the first attempts to introduce natural enemies of insects that were proving troublesome to farmers, an effort that sometimes moved slowly or not at all, but now and again gathered speed and momentum under the impetus of an outstanding success. It had its period of drought when workers in applied entomology, dazzled by the spectacular new insecticides of the 1940's, turned their backs on all biological methods and set foot on "the treadmill of chemical control." But the goal of an insect-free world continued to recede. Now at last, as it has become apparent that the heedless and unrestrained use of chemicals is a greater menace to ourselves than to the targets, the river which is the science of biotic control flows again, fed by new streams of thought.

*   *   *

To some the term microbial insecticide may conjure up pictures of bacterial warfare that would endanger other forms of life. This is not true. In contrast to chemicals, insect pathogens are harmless to all but their intended targets. Dr. Edward Steinhaus, an outstanding authority on insect pathology, has stated emphatically that there is "no authenticated recorded instance of a true insect pathogen having caused an infectious disease in a vertebrate animal either experimentally or in nature." The insect pathogens are so specific that they infect only a small group of insects—sometimes a single species. Biologically they do not belong to the type of organisms that cause disease in higher animals or in plants. Also, as Dr. Steinhaus points out, outbreaks of insect disease in nature always remain confined to insects, affecting neither the host plants nor animals feeding on them.

*   *   *

Through all these new, imaginative, and creative approaches to the problem of sharing our earth with other creatures there runs a constant theme, the awareness that we are dealing with life— with living populations and all their pressures and counterpressures, their surges and recessions. Only by taking account of such life forces and by cautiously seeking to guide them into channels favorable to ourselves can we hope to achieve a reasonable accommodation between the insect hordes and ourselves.

The current vogue for poisons has failed utterly to take into account these most fundamental considerations. As crude a weapon as the cave man's club, the chemical barrage has been hurled against the fabric of life—a fabric on the one hand delicate and destructible, on the other miraculously tough and resilient, and capable of striking back in unexpected ways. These extraordinary capacities of life have been ignored by the practitioners of chemical control who have brought to their task no "high-minded orientation," no humility before the vast forces with which they tamper.

The "control of nature" is a phrase conceived in arrogance, born of the Neanderthal age of biology and philosophy, when it was supposed that nature exists for the convenience of man. The concepts and practices of applied entomology for the most part date from that Stone Age of science. It is our alarming misfortune that so primitive a science has armed itself with the most modern and terrible weapons, and that in turning them against the insects it has also turned them against the earth.

Frank Graham, Jr.
THE RESULTING OUTCRY

When a shortened form of Silent Spring *first appeared in* The New Yorker *in the summer of 1960, several major chemical companies attempted to suppress Rachel Carson's critique of certain pesticides. When the complete book was published in 1962, it provoked a debate which has been compared to the controversy over Harriet Beecher Stowe's* Uncle Tom's Cabin. *Seven years later, in 1969, the environmental debate gained momentum as the questions of population, pollution, urbanization, and conservation simultaneously gained public attention. For a time in 1970, ecological concern even overshadowed other contemporary issues, including the Vietnam war, racial strife, and radicalism to the left and right. This consuming interest in the quality of man's surroundings might never have gained such attention without the persistent presence of Dr. Carson's book. Perhaps her concern and contribution is reflected by two events of 1970: the discovery that mothers' milk contains five times the DDT content allowed for human consumption of cow's milk, and the enactment of severe restrictions upon the use of DDT, with attempts to ban its use entirely. The following selection, from* Since Silent Spring, *by Frank Graham, Jr., field editor of* Audubon *magazine, depicts the crucial early days after Dr. Carson's book first appeared.*

From *Since Silent Spring* by Frank Graham, Jr. Copyright © 1970 by Frank Graham, Jr. Reprinted by permission of the publisher, Houghton Mifflin Company, and Hamish Hamilton.

SILENT SPRING IS NOW NOISY SUMMER. This head appeared over a story in the *New York Times* on July 22, 1962. *Silent Spring* was not yet between hard covers, but the uproar in government, chemical, and agricultural circles was intense. The serialized and abbreviated version of Rachel Carson's book in *The New Yorker* had created a greater stir than anyone earlier had imagined.

The reaction was as varied as it was intense. Friends wrote to Rachel Carson from as far away as the mountains of the Northwest and the villages of the Maritime Provinces that the chief topic of conversation seemed to be *Silent Spring* and chemical pesticides. The Toledo, Ohio, Public Library ordered a large supply of ladybugs from California (at $6.50 a gallon) and released them on its grounds to control aphids. In Bethlehem, Pennsylvania, the *Globe-Times* described the reactions of farm bureau personnel in two nearby counties: "No one in either county farm office who was talked to today had read the book, but all disapproved of it heartily."

Some chemical firms reportedly instructed their scientists to examine the articles line by line, probing for weak spots. The National Agricultural Chemicals Association preferred not to meet Rachel Carson head on. Instead, this powerful lobbyist for the chemical manufacturers expanded its public relations program and published a number of new brochures which reaffirmed the unadulterated blessings of chemical pesticides.

At least one chemical firm preferred to take direct action. On August 2, the Velsicol Chemical Corporation of Chicago addressed a five-page registered letter to Houghton Mifflin, suggesting that the company might wish to reconsider its plans to publish *Silent Spring,* especially in view of the book's "inaccurate and disparaging statements" about chlordane and heptachlor, two chlorinated hydrocarbon pesticides manufactured solely by Velsicol. The letter was signed by Louis A. McLean, Secretary and General Counsel of Velsicol. The letter's sentiments reached their climax in the following paragraph:

> *Unfortunately, in addition to the sincere opinions by natural food faddists, Audubon groups and others, members of the chemical industry in this country and in western Europe must deal with sinister influences, whose attacks on the chemical industry have a dual purpose: (1) to create the false impression that all business is grasping and immoral,*

and (2) to reduce the use of agricultural chemicals in this country and in the countries of western Europe, so that our supply of food will be reduced to east-curtain parity. Many innocent groups are financed and led into attacks on the chemical industry by these sinister parties.

Houghton Mifflin asked for more detailed information on the statements in *Silent Spring* to which Velsicol objected. On receiving the information from McLean, Houghton Mifflin asked an independent toxicologist to review the disputed material. When the toxicologist confirmed the accuracy of Rachel Carson's statements, her publishers informed Velsicol that the book would appear as planned, and nothing further was heard about the matter.

There was furious activity behind the scenes at all levels of government. The Federal Pest Control Review Board (ineffective precursor of the present Federal Committee on Pest Control) met in a troubled session. An observer at that meeting recently recalled its distasteful tone.

"The comments alternated between angry attacks on *Silent Spring* and nasty remarks about Miss Carson," this government official said. "One well-known board member, I recall, said, 'I thought she was a spinster. What's she so worried about genetics for?' Some of the other board members thought this was very funny. I was disgusted by the whole meeting."

Elsewhere the book was taken more seriously. The *Wall Street Journal* had this to say in its August 3 edition about the Department of Agriculture's concern: "Secretary Freeman squelches trigger-happy underlings who itch for a quick rebuttal of Rachel Carson's magazine attacks on safety of chemical insecticides. The Agriculture Department builds a careful defense of its encouragement of insecticide use. An indirect reply: The Department pushes work on non-chemical war against bugs."

Over at the Department of the Interior, Secretary Stewart L. Udall also grasped the significance of the book. He realized that the problem it discussed soon would become a major issue, and he wanted to be able to take an intelligent stand whenever it was called for. Accordingly, he assigned one of his assistants to follow every phase of the book's publication, and the controversy that followed it. (It is interesting to note that an intelligent stand was called for the following year when Udall testified before Senator Abraham Ribicoff's subcommittee which was investigating pesticide use; at that time, Udall became probably the first prominent public servant to point out that we are contaminating the entire environment.)

Congressman John V. Lindsay wrote to Rachel Carson to tell her that he had found *The New Yorker* articles to be "a detailed and persuasive contribution to public awareness of the dangers of our present pest control policy." Lindsay inserted the concluding paragraphs of the first installment in the *Congressional Record,* and told her, "I wish that I could have inserted the entire article."

Part of *Silent Spring*'s impact on the public that summer resulted from the coincidence that *The New Yorker* articles followed almost immediately the thalidomide tragedy, in which it was discovered that this new and inadequately tested chemical tranquilizer, when administered to pregnant women, often caused serious physical abnormalities in their babies. Both the public and government officials, therefore, were receptive to cautionary statements that dealt with other dangerous chemicals. President John F. Kennedy, a regular reader of *The New Yorker,* apparently was familiar with the *Silent Spring* articles that summer. At a White House press conference on August 29, a reporter raised the issue.

"Mr. President," the reporter began, "there appears to be a growing concern among scientists as to the possibility of dangerous long-range side effects from the use of DDT and other pesticides. Have you considered asking the Department of Agriculture or the Public Health Service to take a closer look at this?"

"Yes, and I know they already are," the President replied. "I think particularly, of course, since Miss Carson's book, but they are examining the matter."

Because of the President's concern, the Life Sciences Panel of the President's Science Advisory Committee took over in earnest the study of pesticide use. Its report the following year became one of the milestones in the search for a satisfactory pesticide policy.

These and other reports filtered in to Rachel Carson, now resting at her summer place on the Maine coast. Biologists in the South wrote her to report meetings at local Agricultural Extension Service offices, ostensibly to organize pesticide control committees; in reality, the meetings turned into gabfests between agricultural workers and the representatives of chemical companies about the best means to counteract the message of *Silent Spring.* And from the Netherlands there arrived a letter from Dr. Briejèr. He reported that, even before he had seen *The New Yorker* articles, he had been asked about them by both his country's Minister of Agriculture and its Minister for Foreign Affairs.

"Commercial interests are strong," Briejèr wrote to her. "The use of herbicides is increasing and many complaints about damage are coming in. I am afraid many scientists in the field of plant-protection are on the wrong side."

For the time being, much of the adverse comment about *Silent Spring* was muted. The storm had not yet broken over Rachel Carson's head, as the chemical industry bided its time, waiting for the book's official publication when it could mount its counterattack under the guise of book reviews in both popular and scientific periodicals. Most of the comments now were congratulatory. A physician, writing to *The New Yorker,* remarked that Rachel Carson should feel deep satisfaction if her book received the attention it deserved.

"She will have accomplished as real a service as any physician who devoted a lifetime to patients," he wrote, "and she will have reached a 'practice' encompassing everyone!"

The gratification Rachel Carson experienced from this attention aroused by her book was tempered by dismay that she was now a "controversial figure."

> How dreary to be somebody!
> How public, like a frog
> To tell your name the livelong day
> To an admiring bog!

She shared this horror with Emily Dickinson. West Southport, where she had once felt secure, now seemed to her center stage. On August 1 she wrote to her friend Shirley Briggs in Washington: "So far I have not had to resort to disguises but I am about to have the telephone changed to an unlisted category, and even with that precaution I think I am about to be invaded by the press."

Her book was a major success before it arrived in the bookstores. It had been bought by the Book-of-the-Month Club (with Supreme Court Justice Douglas contributing an article on the book for the club's newsletter). Its advance sale had reached 40,000 copies on September 27, which was publication day. And then the storm broke.

* * *

The most widely distributed anti-Carson article that appeared in any scientific magazine was written by Dr. William J. Darby, a nutritionist at the Vanderbilt University School of Medicine, for *Chemical and Engineering News.* "Her ignorance or bias on some of the considerations throws doubt on her competence to judge policy," Dr. Darby said of Rachel Carson. "For example, she indicates that it is neither wise nor responsible to use pesticides in the control of insect-borne diseases."

Dr. Darby, then, like most of the book's critics, made a great show of refuting statements that Rachel Carson had never made. What she had questioned were the overall methods used in the war against insect-borne diseases; she did not rule out the use of pesticides in all cases.

The chemical industry presented an almost united front against what it considered the menace of Rachel Carson. There were allegations made at the time that certain chemical companies threatened to withdraw their advertising from gardening magazines and newspaper supplements that gave favorable mention to *Silent Spring.* In November, 1962, the Manufacturing Chemists Association began mailing monthly feature stories to news media, stressing the "positive side" of chemical use. Similar material was mailed to about 100,000 individuals. The National Agricultural Chemicals Association doubled its public relations budget. It distributed thousands of copies of reviews that were critical of *Silent Spring.*

This was the gist of the message: "A serious threat to the continued supply of wholesome, nutritious food, and its availability at present-day low prices is manifested in the fear complex building up as a result of recent unfounded, sensational publicity with respect to agricultural chemicals." In the face of even the mildest criticism, the chemical industry has resorted to this theme over and over in the intervening years.

The target of a considerable portion of this material (including "fact kits" which presented the answers wrapped up in the questions) was the medical profession. In the light of this unceasing barrage, it was little wonder that in November the *AMA News* suggested to interested doctors that they contact the chemical trade associations for material to help them answer their patients' questions about pesticides.

Rachel Carson, who was shocked to only a very small degree by the response her book had evoked, nevertheless felt compelled to speak out about this sort of thing. "I can't believe that the AMA seriously believes that an industry with $300 million a year in pesticide sales at stake is an objective source of data on health hazards," she told a reporter.

Meanwhile, an attack began to take shape from what at first appeared to be another quarter. Its source in this case was an organization called The Nutrition Foundation. This organization had been incorporated in 1941 to support fundamental research and education in the science of nutrition. As part of its educational activities early in 1963 it put together a "Fact Kit" on the subject of *Silent Spring.* The kit consisted of a defense of chemical pesticides prepared by the New York State College of Agriculture, and several book reviews that were critical of *Silent Spring.* (One of them was written by I. L. Baldwin, who was the chairman of the National Academy of Sciences-National Research Council's Committee on Pest Control and Wildlife Relationships.) It was accompanied by a letter, written by C. G. King, the president of the Foundation, which stressed the "independence" of those who attacked Rachel Carson's book, and described the book itself as "distorted."

"The problem is magnified," King said, "in that publicists and the author's adherents among the food faddists, health quacks, and special interest groups are promoting her book as if it were scientifically irreproachable and written by a scientist."

A man who could write that sentence obviously should not throw around the word "distorted." King was particularly disturbed because Rachel Carson did not share his admiration for the motives of some leaders in the chemical industry. In his letter, King said he had "known" many of these men, but he could have gone a little further and pointed out their connection with The Nutrition Foundation. The Foundation's membership consisted of fifty-four companies in the food, chemical, and allied industries, many of which depended on chemical pesticides for the cheap production of their raw materials. The presidents of many of these companies served on the Foundation's Board of Trustees. In the other direction, King might have pointed out that the Foundation funneled money, in the form of research grants, from these companies to the nutrition departments of many famous universities. Coincidentally, some of the fiercest attacks on *Silent Spring* came from university nutritionists. As one impartial nutritionist observed: "Where the shot hits, the fur flies."

The Foundation's "Fact Kit" was widely distributed. Included on the mailing lists were university people in many fields, researchers and other staff members of state agricultural experiment stations, the membership of the American Public Health Association, leaders of women's organizations, librarians, visiting nurses, and state, county and municipal officials. The kit was received by tree wardens in Connecticut, by the mayor of a small New Jersey borough, and even by a local Audubon Society in Pennsylvania. Monsanto itself did not operate more efficiently.

Industry supplied the personal touch, too. One of the most active figures on the lecture tour in the months following *Silent Spring*'s publication was Robert H. White-Stevens, assistant director of research and development in the Agricultural Division of the American Cyanamid Company. He made twenty-eight speeches before the end of 1962, extolling the virtues of pesticides and charging that *Silent Spring* was "littered with crass assumptions and gross misinterpretations."

*   *   *

Industry's contention that the scientific world rejected *Silent Spring* was belied by the number of outstanding scientists who declared their admiration for the book. This movement occurred despite the traditional view of the scientist as a man who will not stick his neck out if the cause does not immediately concern him; evidently, many scientists believed that environmental contamination was of immediate concern to them. Robert L. Rudd wrote to Rachel Carson to tell her of the reaction to *Silent Spring* at the University of California.

"I don't have to tell you that *Silent Spring* brought everyone out of the woodwork," Rudd wrote. "The range of opinion in the UC system is as extreme as anywhere else. You'd be surprised, nonetheless, how very much sympathy the points of view we have both held have with some of our specialists in the [Agricultural] Experiment Station and among the scientific disciplines."

Across the country, Loren Eiseley, the celebrated anthropologist at the University of Pennsylvania, added his voice to the support for Rachel Carson. Reviewing her book for the *Saturday Review,* he wrote: "If her present book does not possess the beauty of *The Sea Around Us,* it is because she has courageously chosen, at the height of her powers, to educate us upon a sad, an unpleasant, an unbeautiful topic, and one of our own making. *Silent Spring* should be read by every American who does not want it to be the epitaph of a world not very far beyond us in time."

One of the most balanced reviews of the book was written by LaMont C. Cole [Professor of Ecology at Cornell], in the *Scientific American*. His exhaustive examination of the book allowed him to destroy the heavy-handed attacks on it. Of *Time* magazine's review, he said: "Without citing evidence, (it) proclaims the merits of pesticides in a statement with which, in my opinion, no responsible scientist would want to associate himself."

Whereas the industry-inspired reviews of *Silent Spring* had simply howled indignantly about "distortion," or attacked Rachel Carson for statements she had never made, Cole pinpointed what he believed to be errors in her work. He questioned her statement on page 16 that rotenone and pyrethrum are "simpler inorganic insecticides of prewar days." Pyrethrum, Cole pointed out, is so complicated a substance that "it has frustrated the analysis of the ablest organic chemists," and he suggested that its complicated nature may be "also responsible for frustrating the abilities of insects to develop resistance."

Elsewhere, Cole complained that Rachel Carson had not mentioned the fact the "bees are probably less threatened by modern insecticides than they were by the old arsenicals." He also questioned her statement that resistance among insects causes us to seek ever more potent poisons. The insects that have become resistant to DDT prove less adaptable in other respects, and fall victim to weaker, unrelated pesticides. "I do not for a moment believe," Cole wrote, "that the chemicals are producing super insects."*

The attempt to undermine *Silent Spring* by questioning the facts on which Rachel Carson based her case was doomed to fail. "Errors of fact are so infrequent, trivial and irrelevant to the main theme that it would be ungallant to dwell on them," Cole said. Of Rachel Carson's overall treatment of the use of chemical pesticides, he concluded: "She does not convey an appreciation of the really great difficulties of the problem. . . . But what I interpret as bias and over-simplification may be just what it takes to write a best-seller. . . . If the message of *Silent Spring* is widely enough read and discussed, it may help us toward a much needed reappraisal of current policies and practices."

Silent Spring was an enormous undertaking, as any work is that tries to bring together many disciplines to create a workable synthesis. Rachel Carson saw what most "pest control experts" had not seen—that the specialized view cannot solve the many problems posed by the large-scale use of pesticides. Indeed, such a limited view contributes to the problem. The variety of forms in nature baffles and blinds even scientists, just as the wealth of vegetation in the deep woods shuts off a man's view of all the surrounding

---

* Points which other scientists have raised about inaccuracies in *Silent Spring* include the statement on page 17 that arsenic in chimney soot is the cause of cancer; many, though not all, scientists believe it is the tars in soot that are the chief carcinogens. The role of arsenic, however, as a potent human carcinogen is firmly established. According to Dr. W. C. Hueper of the National Cancer Institute, "The recent severe epidemic of arsenic cancers among German vintners which developed as the result of their use of arsenical pesticides between 1920 and 1940 and which has given rise to the appearance of often multiple and multicentric cancers of the skin, lung, liver, esophagus, larynx, nasal sinuses, etc., should be adequate and ample to convince even the professional sceptics on this point."

Rachel Carson seems to have overstated the threat to robins during the frenetic assault on the elm bark beetle. DDT, used in the control of Dutch elm disease, has taken an enormous toll of robins, but ornithologists see no immediate possibility that the species will plunge "into the night of extinction" (page 105). Certain other species, as we shall see, certainly face extinction.

Perhaps her statement on page 291 that insect pathogens biologically "do not belong to the type of organisms that cause disease in higher animals or in plants" is too general. According to Edward A. Steinhaus of the Center for Pathobiology at the University of California at Irvine, "there may be rare exceptions in which insect pathogens are found to be pathogenic for vertebrate animals; the reported infection of a horse by the fungus *Entomophthora coronata* being a case in point."

forms except those closest to him. It is natural for the specialist to resent the overview. For her temerity, Rachel Carson bore the burden of a great deal of this sort of resentment. The University of California's Robert L. Rudd answered these critics of Rachel Carson when he reviewed *Silent Spring* in *Pacific Discovery* (Nov.–Dec. 1962), the publication of the California Academy of Sciences:

"Are they correct? I should say, 'Yes, in part,' if what is expected is an ultimate knowledge of every aspect of the problem. However, no reviewer, including her critics, has that total knowledge today. . . . In my opinion, she is eminently qualified to present the facts, synthesis and argument she has in *Silent Spring*. I leave it to her critics to do as well."

And we will leave it to Robert Rudd, whose own exhaustive book, *Pesticides and the Living Landscape* (based in large part on his own years of experience in the field and laboratory), appeared shortly afterward, to define the essence of Rachel Carson's book:

"*Silent Spring* is biological warning, social commentary and moral reminder. Insistently, she calls upon technological man to pause and take stock."

III CONTEMPORARY ENVIRONMENTAL PROBLEMS

Paul R. Ehrlich

POPULATION

Paul R. Ehrlich, professor in the department of biological sciences at Stanford University, specializes in population biology. Like a growing number of his colleagues in the biological sciences, he looks upon his training and vocation not with detachment, but as a means to focus attention upon the most immediate crisis that man has forced upon the earth and upon himself. With prophetic fervor and apocalyptic zeal, Dr. Ehrlich has written and lectured widely on the population problem. His concern is based, he says, upon the immediate threat overpopulation brings to "Mankind's Inalienable Rights":

1. *The right to eat well.*
2. *The right to drink pure water.*
3. *The right to breathe clean air.*
4. *The right to decent, uncrowded shelter.*
5. *The right to enjoy natural beauty.*
6. *The right to avoid regimentation.*
7. *The right to avoid pesticide poisoning.*
8. *The right to freedom from thermonuclear war.*
9. *The right to limit families.*
10. *The right to educate our children.*
11. *The right to have grandchildren.*

Dr. Ehrlich is also representative of those biologists who call for new styles of life in American civilization to avoid a steady deterioration of the quality of life. He believes many current crises are closely interrelated, as the above list indicates. Dr. Ehrlich was also influential in the appearance of the "zero-population-growth" movement that gained national attention in 1970.

I have understood the population explosion intellectually for a long time. I came to understand it emotionally one stinking hot night in Delhi a couple of years ago. My wife and daughter and I were returning to our hotel in an ancient taxi. The seats were hopping with fleas. The only functional gear was third. As we crawled through the city, we entered a crowded slum area. The temperature was well over 100, and the air was a haze of dust and smoke. The streets seemed alive with people. People eating, people washing,

From *The Population Bomb* by Paul R. Ehrlich. Copyright 1968 by Paul R. Ehrlich. Reprinted with the permission of the author and Ballantine Books, Inc. Notes to the original have been omitted.

people sleeping. People visiting, arguing, and screaming. People thrusting their hands through the taxi window, begging. People defecating and urinating. People clinging to buses. People herding animals. People, people, people, people. As we moved slowly through the mob, hand horn squawking, the dust, noise, heat, and cooking fires gave the scene a hellish aspect. Would we ever get to our hotel? All three of us were frankly, frightened. It seemed that anything could happen—but, of course, nothing did. Old India hands will laugh at our reaction. We were just some overprivileged tourists, unaccustomed to the sights and sounds of India. Perhaps, but since that night I've known the *feel* of overpopulation.

*   *   *

It has been estimated that the human population of 6000 B.C. was about five million people, taking perhaps one million years to get there from two and a half million. The population did not reach 500 million until almost 8,000 years later—about 1650 A.D. This means it doubled roughly once every thousand years or so. It reached a billion people around 1850, doubling in some 200 years. It took only 80 years or so for the next doubling, as the population reached two billion around 1930. We have not completed the next doubling to four billion yet, but we now have well over three billion people. The doubling time at present seems to be about 37 years. Quite a reduction in doubling times: 1,000,000 years, 1,000 years, 200 years, 80 years, 37 years. Perhaps the meaning of a doubling time of around 37 years is best brought home by a theoretical exercise. Let's examine what might happen on the absurd assumption that the population continued to double every 37 years into the indefinite future.

If growth continued at that rate for about 900 years, there would be some 60,000,000,000,000,000 people on the face of the earth. Sixty million billion people. This is about 100 persons for each square yard of the Earth's surface, land and sea. A British physicist, J. H. Fremlin, guessed that such a multitude might be housed in a continuous 2,000-story building covering our entire planet. The upper 1,000 stories would contain only the apparatus for running this gigantic warren. Ducts, pipes, wires, elevator shafts, etc., would occupy about half of the space in the bottom 1,000 stories. This would leave three or four yards of floor space for each person. I will leave to your imagination the physical details of existence in this ant heap, except to point out that all would not be black. Probably each person would be limited in his travel. Perhaps he could take elevators through all 1,000 residential stories but could travel only within a circle of a few hundred yards' radius on any floor. This would permit, however, each person to choose his friends from among some ten million people! And, as Fremlin points out, entertainment on the worldwide TV should be excellent, for at any time "one could expect some ten million Shakespeares and rather more Beatles to be alive."

Could growth of the human population of the Earth continue beyond that point? Not according to Fremlin. We would have reached a "heat limit." People themselves, as well as their activities, convert other forms of energy into heat which must be dissipated. In order to permit this excess heat to radiate directly from the top of the "world building" directly into space, the atmosphere would have been pumped into flasks under the sea well before the limiting population size was reached. The precise limit would depend on the technology of the day. At a population size of one billion billion people, the temperature of the "world roof" would be kept around the melting point of iron to radiate away the human heat generated.

But, you say, surely Science (with a capital "S") will find a way for us to occupy the other planets of our solar system and eventually of other stars before we get all that crowded. Skip for a moment the virtual certainty that those planets are uninhabitable. Forget also the insurmountable logistic problems of moving billions of people off the Earth. Fremlin has made some interesting calculations on how much time we could buy by occupying the planets of the solar system. For instance, at any given time it would take only about 50 years to populate Venus, Mercury, Mars, the moon, and the moons of Jupiter and Saturn to the same population density as Earth.

What if the fantastic problems of reaching and colonizing the other planets of the solar system, such as Jupiter and Uranus, can be solved? It would take only about 200 years to fill them "Earth-full." So we could perhaps gain 250 years of time for population growth in the solar system after we had reached an absolute limit on Earth. What then? We can't ship our surplus to the stars. Professor Garrett Hardin of the University of California at Santa Barbara has dealt effectively with this fantasy. Using extremely optimistic assumptions, he has calculated that Americans, by cutting their standard of living down to 18 percent of its present level, could in *one year* set aside enough capital to finance the exportation to the stars of *one day's* increase in the population of the world.

Interstellar transport for surplus people presents an amusing prospect. Since the ships would take generations to reach most stars, the only people who could be transported would be those willing to exercise strict birth control. Population explosions on space ships would be disastrous. Thus we would have to export our responsible people, leaving the irresponsible at home on Earth to breed.

Enough of fantasy. Hopefully, you are convinced that the population will have to stop growing sooner or later and that the extremely remote possibility of expanding into outer space offers no escape from the laws of population growth. If you still want to hope for the stars, just remember that, at the current growth rate, in a few thousand years everything in the visible universe would be converted into people, and the ball of people would be expanding with the speed of light! Unfortunately, even 900 years is much too far in the future for those of us concerned with the population explosion. As you shall see, the next *nine* years will probably tell the story.

*   *   *

How did we get into this bind? It all happened a long time ago, and the story involves the process of natural selection, the development of culture, and man's swollen head. The essence of success in evolution is reproduction. Indeed, natural selection is simply defined as differential reproduction of genetic types. That is, if people with blue eyes have more children on the average than those with brown eyes, natural selection is occurring. More genes for blue eyes will be passed on to the next generation than will genes for brown eyes. Should this continue, the population will have progressively larger and larger proportions of blue-eyed people. This differential reproduction of genetic types is the driving force of evolution; it has been driving evolution for billions of years. Whatever types produced more offspring became the common types. Virtually all populations contain very many different genetic types (for reasons that need not concern us), and some are always outreproducing others. As I said, reproduction is the key to winning the evolutionary game. Any structure, physiological process, or pattern of behavior that leads to greater reproductive success will tend to be perpetuated. The entire process by which man developed involves thousands of millenia of our ancestors being more successful breeders than their relatives. Facet number one of our bind—the urge to reproduce has been fixed in us by billions of years of evolution.

Of course through all those years of evolution, our ancestors were fighting a continual battle to keep the birth rate ahead of the death rate. That they were successful is attested to by our very existence, for, if the death rate had overtaken the birth rate for any substantial period of time, the evolutionary line leading to man would have gone extinct. Among our apelike ancestors, a few million years ago, it was still very difficult for a mother to rear her children successfully. Most of the offspring died before they reached reproductive age. The death rate was near the birth rate. Then another factor entered the picture—cultural evolution was added to biological evolution.

Culture can be loosely defined as the body of nongenetic information which people pass from generation to generation. It is the accumulated knowledge that, in the old days, was passed on entirely by word of mouth, painting, and demonstration. Several thousand years ago the written word was added to the means of cultural transmission. Today culture is passed on in these ways, and also through television, computer tapes, motion pictures, records, blueprints, and other media. Culture is all the information man possesses except for that which is stored in the chemical language of his genes.

The large size of the human brain evolved in response to the development of cultural information. A big brain is an advantage when dealing with such information. Big-brained individuals were able to deal more successfully with the culture of their group. They were thus more successful reproductively than their smaller-brained relatives. They passed on their genes for big brains to their numerous offspring. They also added to the accumulating store of cultural information, increasing slightly the premium placed on brain size in the next generation. A self-reinforcing selective trend developed— a trend toward increased brain size.

But there was, quite literally, a rub. Babies had bigger and bigger heads. There were limits to how large a woman's pelvis could conveniently become. To make a long story short, the strategy of evolution was not to make a woman bell-shaped and relatively immobile, but to accept the problem of having babies who were helpless for a long period while their brains grew after birth. How could the mother defend and care for her infant during its unusually long period of helplessness? She couldn't, unless Papa hung around. The girls are still working on that problem, but an essential step was to get rid of the short, well-defined breeding season characteristic of most mammals. The year-round sexuality of the human female, the long period of infant dependence on the female, the evolution of the family group, all are at the roots of our present problem. They are essential ingredients in the vast social phenomenon that we call sex. Sex is not simply an act leading to the production of offspring. It is a varied and complex cultural phenomenon penetrating into all aspects of our lives—one involving our self-esteem, our choice of friends, cars, and leaders. It is tightly interwoven with our mythologies and history. Sex in man is necessary for the production of young, but it also evolved to ensure their successful rearing. Facet number two of our bind—our urge to reproduce is hopelessly entwined with most of our other urges.

Of course, in the early days the whole system did not prevent a very high mortality among the young, as well as among the older members of the group. Hunting and food-gathering is a risky business. Cavemen had to throw very impressive cave bears out of their caves before the men could move in. Witch doctors and shamans had a less than perfect record at treating wounds and curing disease. Life was short, if not sweet. Man's total population size doubtless increased slowly but steadily as human populations expanded out of the African cradle of our species.

Then about 8,000 years ago a major change occurred—the agricultural revolution. People began to give up hunting food and settled down to grow it. Suddenly some of the risk was removed from life. The chances of dying of starvation diminished greatly in some human groups. Other threats associated with the nomadic life were also reduced, perhaps balanced by new threats of disease and large-scale warfare associated with the development of cities. But the overall result was a more secure existence than before, and the human population grew more rapidly. Around 1800, when the standard of living in what are today the DCs [Developed Countries] was dramatically increasing due to industrialization, population growth really began to accelerate. The development of medical science was the straw that broke the camel's back. While lowering death rates in the DCs was due in part to other factors, there is no question that "instant death control," exported by the DCs, has been responsible for the drastic lowering of death rates in the UDCs [Undeveloped Countries]. Medical science, with its efficient public health programs, has been able to depress the death rate with astonishing rapidity and at the same time drastically increase the birth rate; healthier people have more babies.

* * *

The key to the whole business, in my opinion, is held by the United States. We are the most influential superpower; we are the richest nation in the world. At the same time we are also just one country on an ever-shrinking planet. It is obvious that we cannot exist unaffected by the fate of our fellows on the other end of the good ship Earth. If their end of the ship sinks, we shall at the very least have to put up with the spectacle of their drowning and listen to their screams. Communications satellites guarantee that we will be treated to the sights and sounds of mass starvation on the evening news, just as we now can see Viet Cong corpses being disposed of in living color and listen to the groans of our own wounded. We're unlikely, however, to get off with just our appetites spoiled and our consciences disturbed. We are going to be sitting on top of the only food surpluses available for distribution, and those surpluses will not be large. In addition, it is not unreasonable to expect our level of affluence to continue to increase over the next few years as the situation in the rest of the world grows ever more desperate. Can we guess what effect this growing disparity will have on our "shipmates" in the UDCs? Will they starve gracefully, without rocking the boat? Or will they attempt to overwhelm us in order to get what they consider to be their fair share?

We, of course, cannot remain affluent and isolated. At the moment the United States uses well over half of all the raw materials consumed each year. Think of it. Less than 1/15th of the population of the world requires more than all the rest to maintain its inflated position. If present trends continue, in 20 years we will be much less than 1/15th of the population, and yet we may use some 80 percent of the resources consumed. Our affluence depends heavily on many different kinds of imports: ferroalloys (metals used to make various kinds of steel), tin, bauxite (aluminum ore), rubber, and so forth. Will other countries, many of them in the grip of starvation and anarchy, still happily supply these materials to a nation that cannot give them food? Even the technological optimists don't think we can free ourselves of the need for imports in the near future, so we're going to be up against it. But, then, at least our balance of payments should improve!

So, beside our own serious population problem at home, we are intimately involved in the world crisis. We are involved through our import-export situation. We are involved because of the possibilities of global ecological catastrophe, of global pestilence, and of global thermonuclear war. Also, we are involved because of the humanitarian feelings of most Americans.

We are going to face some extremely difficult but unavoidable decisions. By how much, and at what environmental risk, should we increase our food production in an attempt to feed the starving? How much should we reduce the grain-finishing of beef in order to have more food for export? How will we react when asked to balance the lives of a million Latin Americans against, say, a 30 cent per pound rise in the average price of beef? Will we be willing to slaughter our dogs and cats in order to divert pet food protein to the starving masses in Asia? If these choices are presented one at a time, out of context, I predict that our behavior will be "selfish." Men do not seem to be able to focus emotionally on distant or long-term events. Immediacy seems to be necessary to elicit "selfless" responses. Few Americans could sit in the same room with a child and watch it starve to death. But the death of several million children this year from starvation is a distant, impersonal, hard-to-grasp event. You will note that I put quotes around "selfish" and "selfless." The words describe the behavior only out of context. The "selfless" actions necessary to aid the rest of the world and stabilize the population are our only hope for survival. The "selfish" ones work only toward our destruction. Ways must be found to bring home to all the American people the reality of the threat to their way of life—indeed to their very lives.

Obviously our first step must be to immediately establish and advertise drastic policies designed to bring our own population size under control. We must define a goal of a stable optimum population size for the United States and display our determination to move rapidly toward that goal. Such a move does two things at once. It improves our chances of obtaining the kind of country and society we all want, and it sets an example for the world. The second step is very important, as we also are going to have to adopt some very tough foreign policy positions relative to population control, and we must do it from a psychologically strong position. We will want to disarm one group of opponents at the outset: those who claim that we wish others to stop breeding while we go merrily ahead. We want our propaganda based on "do as we do"—not "do as we say."

* * *

I wish I could offer you some sugarcoated solutions, but I'm afraid the time for them is long gone. A cancer is an uncontrolled multiplication of cells; the population explosion is an uncontrolled multiplication of people. Treating only the symptoms of cancer may make the victim more comfortable at first, but eventually he dies— often horribly. A similar fate awaits a world with a population explosion if only the symptoms are treated. We must shift our efforts from treatment of the symptoms to the cutting out of the cancer. The operation will demand many apparently brutal and heartless decisions. The pain may be intense. But the disease is so far advanced that only with radical surgery does the patient have a chance of survival.

So far I have talked primarily about the strategy for easing us through the hazardous times ahead. But what of our ultimate goals? That, of course, is something that needs a great deal of discussion in the United States and elsewhere. Obviously, we need a stable world population with its size rationally controlled by society. But what should the size of that population be? What is the optimum number of human beings that the Earth can support? This is an extremely complex question. It involves value judgments about how crowded we should be. It also includes technical questions of how crowded we *can* be. Research should obviously be initiated in both areas immediately.

If we are to decide how crowded we should be, we must know a great deal more about man's perception of crowding and about how crowding affects human beings. Certainly people in different cultures and subcultures have different views of what densities of people (people per unit area) constitute crowding under different conditions. But what exactly are those densities and conditions? Under what conditions do people consider themselves neither crowded nor lonely? Research on these questions has barely been started. It must be accompanied by studies of how crowding affects people, including both "overcrowding" (too many people per unit area) and "undercrowding" (too few per unit area). These problems are more difficult to study, especially since the effects of crowding are often confounded by poverty, poor diet, unattractive surroundings, and other related phenomena.

But difficult as these problems are, they must be investigated. We know all too well that when rats or other animals are overcrowded, the results are pronounced and usually unpleasant. Social systems may break down, cannibalism may occur, breeding may cease altogether. The results do not bode well for human beings as they get more and more crowded. But extrapolating from the behavior of rats to the behavior of human beings is much more risky than extrapolating from the physiology of rats to the physiology of human beings. Man's physical characteristics are much more ratlike than are his social systems. This research must be done on man.

Within the limits imposed by nature, I would view an optimum population size for the Earth to be one permitting any individual to be as crowded or as alone as he or she wished. Enough people should be present so that large cities are possible, but people should not be so numerous as to prevent people who so desire from being hermits. Pretty idealistic, but not impossible in theory. Besides, some pretty far-reaching changes are going to be required in human society over the next few decades, regardless of whether or not we stop the population explosion. We've already reached a density at which many of our institutions no longer function properly. As the distinguished historian, Walter Prescott Webb, pointed out 16 years ago, with the closing of the World Frontier, a set of basic institutions and attitudes became outdated. When the Western Hemisphere was opened to exploitation by Europeans, a crowded condition suddenly was converted into an uncrowded one. In 1500 the ratio of people to available land in Europe was estimated to have been about 27 people per square mile. The addition of the vast, virtually unpopulated frontier of the New World moved this ratio back down to less than five per square mile. As Webb said, the frontier was, in essence, "a vast body of wealth without proprietors." Europeans moved rapidly to exploit the spatial, mineral, and other material wealth of the New World. They created an unprecedented economic boom that lasted some 400 years. The boom is clearly over, however, at least as far as land is concerned. The man/land ratio went beyond 27 people per square mile again before 1930. Since all of the material things on which the boom depended also come ultimately from the land, the entire boom is also clearly limited. Of course, how to end that boom gracefully, without the most fantastic "bust" of all time, is what this book is all about.

Somehow we've got to change from a growth-oriented, exploitative system to one focused on stability and conservation. Our entire system of orienting to nature must undergo a revolution. And that revolution is going to be extremely difficult to pull off.

John Burchard
URBAN SOCIETY

Until ecology drew national attention in 1969–1970, it was a term which defined specializations in biology (natural sciences) and urban sociology

From "The Culture of Urban America," by John Burchard, in *Environment and Change: The Next Fifty Years,* edited by William R. Ewald, Jr. Copyright © 1968 by Indiana University Press. Reprinted by permission.

(social sciences). The vision and criticism of American life that emerged out of sociology, architecture and engineering complemented the biological image. The following selection is an example of environmental inquiry in the framework of urban planning. It is by John Burchard, architectural historian, environmental designer, and professor of environmental design at Berkeley. Burchard presented the essay at the Fiftieth Year Consultation of the American Institute of Planners, held in 1966. The paper is self-consciously humanistic, and developed the theme of the Consultation—"Optimum Environment with Man as the Measure." Urban planners like Burchard point to a growing crisis, since America is an urban nation, with 80 percent of its population in metropolitan areas. They argue that the future of the cities—good or bad—will determine the future of America, and probably of mankind and civilization. Burchard proposes some intriguing tests for the quality of life in several major cities.

. . . I am convinced that for all such people and, indeed, for all of us, the city is meaningless and even menacing unless it permits—or offers—pleasure throughout life. The pleasure is to be positive, not mere absence of pain. The enjoyment should be relatively unabandoned, not all premeditated and intellectual. The thing enjoyed may be very simple or very esoteric, from a bird call to a symphony, from a bill board to a Leonardo, from peanut butter to pressed duck, from something that happens every day, like my good breakfast, to a performance of *Traviata* by Tebaldi in her prime, such as I shall never encounter again.

Much of this pleasure must and should be low-keyed, and some of the pleasures should be those of routine and some those of accident. How have I slept? Was the night noisy or quiet? Did I have to use sleeping pills? Or Flents? What did the air feel like as it brushed my cheeks at dawn? Or did it brush me at all? Were the birds awake a little before me and did their chorus help to bring me to consciousness? Or were there no birds at all? Did I see something charming when my eyes first opened? Or only a broken window pane across a dirty street? Could I feel the sun on my back? Or was there no sun? Was it a pleasure to wash? Was it a joy to don the clothes I donned and to see myself in the morning mirror? Was breakfast a positive pleasure or a standard routine of two items whose merit was that they had no distinctive flavor but were supposed to contain the right amount of vitamins? When I left my home, did I walk or ride in pleasant quarters, among trees and verdure, or only in dingy streets;

was my public conveyance clean, safe, even pleasant, the driver courteous? Did I see anything amusing on the way? Was my work a place to admire? Did I feel like singing at my work, or was the coffee break the only anticipated activity of the day?

Certainly I do not need to plough on until Morpheus kisses me good night. What this much enumeration signifies must be clear. All of our cities should offer all of these contentments—these simple things. As it is, no city offers all of them for any of its people all of the time; some cities offer some of them for some of the people some of the time; but in every city there are many people for whom none of them is offered any of the time.

Beyond these gentle pleasures, some larger ones are available in some cities, American and foreign. There are, for example, the countless pleasures of various neighborhood "do's," organized or spontaneous. Beyond these there are the specialized and grander pleasures we associate with "great cities."

I think it undebatable that one set of the "grander" pleasures involves what is often called the high culture. Of course there are other sources of urban joy: bird-watching, street dancing, hot-rodding, fishing in the Seine, going to a rally in Trafalgar Square, quiet pleasures and loud ones, gentle ones and vulgar ones. It would not be right to say that the pleasures of the high culture are the most important, but it would be equally absurd to say they are not important at all. I shall say a little more about them because I know more about them than I do about street fun and games, and because I have had a chance to savor them in many parts of the world.

The first and perhaps the only thing to be said about them is that in the present state of our knowledge, it is quite impossible to quantify them to anyone's satisfaction or to crank them into an urban cost-benefit analysis. I made the attached table of Urban Amenity with precisely that in mind. I began by drawing up a list of 24 things generally of the high culture (in my book as in the Greek book, big sports can easily be more important than little theater). Of course it is a highly personal and subjective list. Of course the items may not be of equal weight; there may be duplications; there may be inconsistencies. Then I drew up a list of sixteen cities, eight foreign, eight domestic, with which I had more than a casual familiarity. All of these cities I regard as positively pleasurable, viewed only from the standpoint of more or less traditional culture. This does not mean I can think of no others which *might* be individually more pleasurable than some one on the list (e.g., Lisbon, Mexico City, Kyoto, Melbourne, and maybe Philadelphia). Save that I included no "bad" cities, the list made no pretensions to being decisive. The presence of any city on the list did mean that I thought well of it; the absence of any city did not mean that I thought ill of it; nor was there any implication that this particular foreign set were the best eight foreign cities, or this particular American set the best eight American cities. It would have involved only labor to extend the list as far as one wanted to include all cities, but the thing I sought to show would not have justified it.

Having established this faulty matrix, I put a 1, a ½, and a 0 opposite each characteristic and simply totalled the subjective results. I need no Ph.D. candidate to tell me this was terrible technique. I would only say that the nature of the problem is so subjective that more sophisticated techniques would be unlikely to produce more sophisticated answers. Then I published the results. The consequence of this confirmed my purpose, though it produced more angry correspondence than I craved. How could I really have left out City X? Did I really think Boston was "better" than Philadelphia? (Yes) Had I ever been to Mexico City? (Yes) Do I know how Venice smells at low tide on a hot day? (Yes) Do I really think Pittsburgh's *site* is more spectacular than Rome's? (Yes) How could I say there are no great restaurants in Washington? (I can and do.) What was good about it was that so many people seemed to care. What was also good was that the correspondence confirmed my convictions about the qualitative nature of the whole affair. What was bad was that scholars seemed quite as prone as the journalists to go off halfcocked.

As things turned out in the table—for which, it must by now be evident, I claim not even personal, let alone consensual validity—I probably would have rated the factors in a somewhat different order than they came out. On the other hand, I did not need to make the table to learn that, measured by this particular set of characteristics and on an Occidental scale, Paris, London, Rome, and New York were the greatest cities in our present world.

It should be clear also that this rubric is not enough even if it could be soberly and reliably constructed. It does not even suggest that the "greatest" cities would always be the ones in which it might be the most pleasant to live. I myself, for example, have chosen to live the best part of fifty years in Boston and San Francisco, which come out only in the middle of this score card. What is not so certain is whether this could have been the choice had Paris, Rome, London, and New York not existed and been reasonably accessible at that.

It would be interesting to see more serious research on quantifying the amenities. I doubt that the results will be qualitatively convincing. As long as they are not, there is a very serious risk that hard-headed but unhedonistic planners will leave them out of the urban calculation altogether.

As I have remarked elsewhere, there is no certainty that urban beauty spots may offer any solace to the discontented. Would the gardens of the Tuileries have satisfied the sansculottes of the Faubourg St. Antoine had they been public instead of royal gardens? It is to be doubted. There is even the chance that, given enough anger, the rioters might turn against the city's beauty as a better symbol of the target of the wrath than their own districts can ever be. But these seem to me risks that must be run. In our urgent solicitude to make our cities reasonable places for 10 percent of us, we must not neglect the other 90 percent.

Beyond delectability, there may, rarely, be ecstasy. Ecstasy is not something to be desired too much of the time; nor is it long prolonged. One needs always to ask whose ecstasy, and to remember that it cannot be charted in advance. It is impossible to predict the moment when one's heart will stand still, and I am not sure that planners can do anything to increase the probability of an ecstatic experience.

But the other urban amenities can be planned for and struggled for. They may be beautiful or just fun. None of the traditional urban beauties is really obsolete. Most of these are the result of past planning and architecture. The retreat from the Grand Plan, mainly on ideological grounds, created an antagonism to formal beauty that has not yet died away and that has mortgaged the future very heavily, since the great spaces of the older cities will seldom be duplicated now; new ones are not proliferating, and the inexorable march of the urban numbers means that each we have, like the parks

Urban Amenity Score Sheet

	Paris	Rome	London	New York	Stockholm	Chicago	Boston	Rio	San Francisco	Sydney	Venice	Washington	Istanbul	Pittsburgh	Los Angeles	Dallas/Ft. Worth	Totals
Fine river, lake, etc.	1	1	1	1	1	1	1	1	1	1	1	½	1	1	0	0	13½
Great park(s)	1	1	1	1	1	1	½	1	1	1	0	1	0	1	½	1	13
Distinguished buildings	1	1	1	1	½	1	1	1	½	½	1	1	1	½	½	½	13
Distinguished museum(s)	1	1	1	1	½	1	1	½	0	0	1	1	1	1	1	½	12½
Readable plan	1	½	1	1	1	1	½	1	1	½	1	1	½	1	½	0	12½
Great university	1	½	1	1	1	1	1	0	1	1	0	½	1	1	1	½	12½
Diverse neighborhoods	1	1	1	1	½	½	½	1	1	½	1	½	½	½	½	0	11
Great eating	1	1	½	1	½	½	½	1	1	½	1	½	½	0	1	½	11
Fine music	½	1	1	1	½	1	1	½	1	½	0	½	0	1	1	½	11
General boscage	1	1	1	½	1	½	1	1	½	1	0	1	0	0	½	½	10½
Glamorous site	½	½	½	1	1	½	½	1	1	1	1	0	1	1	0	0	10½
Great sports	1	1	1	1	0	1	1	1	1	1	0	0	0	0	1	½	10½
Great avenue(s)	1	½	1	1	½	1	½	1	0	½	1	1	½	0	0	0	9½
Fine squares	1	1	1	½	1	0	½	0	½	½	1	0	½	½	0	0	8
Important visible past	1	1	1	1	½	0	1	0	0	0	1	½	1	0	0	0	8
Good air	0	1	0	0	1	0	0	1	1	1	1	0	1	0	0	½	7½
Fine libraries	1	1	1	1	0	1	1	0	0	0	0	1	0	0	0	0	7
Exciting shop windows	1	½	1	1	0	½	0	½	0	0	0	0	0	0	½	0	6
																1	
Generally pleasant climate	½	1	0	½	½	0	½	½	1	1	0	0	0	0	½	0	6
Fountains	1	1	1	0	1	0	1	0	0	0	0	0	0	0	0	0	5
Theater	1	1	1	1	½	0	0	0	0	0	0	0	0	0	0	½	5
Art in the streets	1	1	0	0	1	0	0	0	½	0	1	0	0	0	0	0	4½
Private galleries	1	½	1	1	0	½	½	0	0	0	0	0	0	0	0	0	4½
Many opportunities for participatory recreation	0	0	0	½	1	½	½	½	0	1	0	0	0	0	0	0	4
Totals	20½	20	19	19	15½	14½	14	13½	13	12½	12	10	9½	9	8	6½	

and the wilderness, are desperately striving to serve more people without being drowned. The planner's traditional aversion to beauty is not something to look back upon with pride.

* * *

Everything we are talking about here is no more than amiable conversation so long as funds and manpower are not available to cities in really large amounts; and so long as so much of the scientific and technological resource of the country is engaged elsewhere. . . .

It is probably true that we tend to see some other cities through more rosy glasses than we view our own (unless we persist in believing we can learn nothing anywhere else—"I wouldn't trade Shreveport for Rome any day"). This is no doubt because we are tourists with more time and even more daily money to spend on urban joy than we will allow ourselves at home.

Happy the citizen whose city is so abounding in diverse opportunities for personal pleasure that he has never quite explored them all and thus can approach them in some senses as a tourist. But the things one is used to matter too. One can get used to anything. It is only a matter of faith that it is better to get used to a thrush than to a pigeon. To make this possible is a task, not for narrow-nosed reformers, but for planners with sympathy. They must not think it pandering to open opportunities for activities which they do not understand, or, if understanding, do not approve. The computers may provide such men with some information they need, but the projections have to come from not only the lucubrations of the mind, but also from the beatings of the heart.

Robert Reinow and Leona Train Reinow
POLLUTION

*Robert Reinow and Leona Train Reinow have been observers of the growing
environmental crisis since the 1930's. Dr. Reinow is professor of political
science at the State University of New York at Albany. He and his wife are
representative of environmental spokesmen whose professional interests lie
elsewhere, but who have been drawn into the debate. They view the question
of man in the environment through the eyes of the humanities and social
sciences, but as informed by the biological sciences. Moment in the Sun,
subtitled A Report on the Deteriorating Quality of the American Environment,
published first in 1967, reappeared in 1969 to influence strongly the steps
leading to the national Earth Day on April 26, 1970. The Reinows call for
less splintered specialization into the different crises of American life, and
propose a new reunion between the "two cultures" of sciences and humanities. Effective action demands, they believe, the ability to "be prepared to
look through a telescope as often as . . . through a microscope." Ecology
may be the missing link for this new integration, and the alternatives are too
forbidding to accept. The following selection is an entire chapter, "38
Cigarettes a Day," in their book.*

A recent scientific analysis of New York City's atmosphere concluded that a New Yorker on the street took into his lungs the equivalent in toxic materials of 38 cigarettes a day. Suddenly all the scientific jargon, official warnings, reams of statistics—the overwhelming
avalanche of damning facts concerning America's air pollution—took
focus. Here was a reduction of the tons of soot, sulphides, monoxide,
hydrocarbons, etc., into simple, understandable, personal terms.

These figures are of vital interest to two-thirds of the population,
which is the percentage of Americans who already live in 212 standard metropolitan areas having only 9 percent of the nation's land
area but 99 percent of its pollution. Some cities outdo Manhattan
and on days of cloud and atmospheric inversion actually kill off small
segments of their excess population (involuntarily, of course).

Smog production seems to be a cooperative effort among our
great cities. "A great deal of the smoke and dirty air in New York
City comes each morning from the industrial areas of New Jersey,"

Reprinted from *Moment in the Sun* by Robert Reinow and Leona Train Reinow.
Copyright © 1967 by Robert Reinow and Leona Train Reinow and used by permission of the publisher, The Dial Press and Paul R. Reynolds, Inc. Notes to the
original have been omitted.

accused former Mayor Wagner as, smiling apologetically, he testified at an October, 1964, hearing on air pollution. Then his innate fairness forced him to add: "We return the compliment each afternoon, depending on the prevailing winds, or we pass some of our smoke and gases on to our Long Island or Connecticut neighbors."

But, definitely, New York was getting the worst of everything. It is "the terminus of a 3,000-mile-long sewer of atmospheric filth starting as far away as California and growing like a dirty snowball all the way." New Jersey's champion, Chairman William Bradley of the state's pollution control commission, felt that it was New Jersey who was really behind the eight ball or, rather, dirty snowball. "We feel incapable of coping," he sighed dramatically.

What, actually, is it we are talking about when we rant about air pollution? What it is varies from city to city and from industrial complex to industrial complex. There is a conglomeration of particles—bits of metal, the metallic oxides, tar, stone, carbon, and ash, the aerosols, mists of oils, and all manner of soot.

Strangely enough, Anchorage, Alaska, in the clear and pristine North, excels in this air filth, with Charleston, West Virginia, East Chicago, Phoenix, and Los Angeles treading eagerly on its heels. The electrostatic precipitator in a factory chimney or the use of a whirling water bath for smoke emissions can remove many of these solids, but such devices cost money, and money (and the treasuring thereof) is still our number one consideration.

Much more serious than filthy particles is the sulphur dioxide that comes from the combustion of all heavy fuel oils, coal, and coke. We have visual evidence that this substance eats away brick, stone, and metal bridges, but we have not taken time off to discover what it does to the human lung. A derivative, sulphur trioxide, is the common sulphuric acid, which we know eats into the lungs, eyes, and skin, but again research as to how to extirpate it from the air we breathe does not add appreciably to the GNP.

Technology's contribution to the air we breathe includes—in addition to the particulates, aerosols, and sulphur oxides—a whole legion of grisly gases, among them, carbon dioxide, carbon monoxide, hydrofluoric acid, hydrochloric acid, ammonia, organic solvents, aromatic benzypyrene, deadly ozone, and perhaps another 500 or more lethal emissions (some day we shall have discovered thousands).

A few years ago the oil refineries and factories were assigned most of the blame for city smog; later, incinerators took the abuse; at length it was demonstrated beyond a doubt that from 60 to 85 percent of most city smog is caused by man's best friend, the effusive automobile.

Such smog, strangely enough, has lately been discovered to be the result of the action of sunlight on the incompletely combusted automobile exhaust gases, mainly carbon monoxide, the hydrocarbons, and nitrogen oxides. An unbelievably complex and varied "mishmash" of photochemical reactions takes place all day long from dawn to dark. Out of this witches' cauldron, whose catalyst is the sunshine, emerges a whole army of killing compounds: olefins (synergetic hydrocarbons "hungry to react to something"), ketene, peroxyacetylnitrate, sulphuric acid, aldehydes, and, probably most vicious of all, ozone. The automobile is a versatile chemical factory that can produce almost anything you might wish to dial. Of all these perverse and malicious agents, man knows as yet almost nothing about what they do to humans over a period of time.

Nor is there much evidence that he greatly cares.

Senator Edmund S. Muskie of Maine, who has interested himself deeply in the pure air campaign and the Clean Air Act of 1963, says that, regarding air pollution, we are as ignorant of the components and what to do about them as we were about water pollution fifty years ago. There are about eight of the possibly thousands of constituents of automobile exhaust whose presence and amounts we have documented. According to a competent study the automobile emits into the atmosphere for each 1,000 gallons of gasoline consumed:

carbon monoxide	3,200 pounds
organic vapors	200–400 pounds
oxides of nitrogen	20–75 pounds
aldehydes	18 pounds
sulphur compounds	17 pounds
organic acids	2 pounds
ammonia	2 pounds
solids (zinc, metallic oxides, carbon)	.3 pounds

Now, much has been made of the sulphuric acid fumes that burn the eyes and make holes in ladies' nylons. Lately, pushed by such indomitable souls as Senator Muskie, the unburned cancer-causing hydrocarbons (organic vapors) have been getting their share of attention in the mandatory action taken in smog-bound California. Thus, it was California who led the way for the imposition, in March, 1966, of limitations on the amounts of carbon monoxide and hydrocarbons that may emerge from automobile exhaust pipes. The standards, to take effect on the 1968 models, both of domestic and imported large cars, were to raise the price tags on new models not more than $45.00.

Sighing, an industry executive admitted that the automobile manufacturers had not the ghost of a chance of evading the afterburner hassle any longer. "Politicians, it is quite clear, have come to regard this issue as they do home and mother," he said. All would seem to be set for a cleaner, more wholesome atmosphere in the cities where the automobile is king.

But now a sinister new note has been sounded that suddenly cloaks the whole campaign against the hydrocarbons and the affable acquiescence of the auto manufacturers with the faint smell of herring. Why are we so cavalierly overlooking the nitrogen oxides, perhaps the most dangerous family of all produced by internal combustion, which will not be affected by any afterburner or catalytic muffler ever made? In our first efforts to eliminate a portion of the smog problem have we actually done no more than introduce a placebo that will lull the public into a false and fatal reassurance?

Nitrogen oxides are formed in all combustion processes; the greater the pressure, the greater the amount of them. If we eliminate the hydrocarbons with which they react, what will happen? Will they then turn on us and cause even greater havoc than at present —perhaps greater havoc than the hydrocarbons and all their relatives together? We know that irritations of the mucous membranes would greatly increase. And while hydrocarbons build up in the body for the kill, nitrogen oxides like to do a faster, neater job.

Yet we are ignoring them. The truth is, the complicated and expensive accessory needed to deal with the nitrogen oxides would be almost exactly opposite to that necessary for dealing with the hydrocarbons. An automobile properly equipped to get rid of smog effectively would probably be an elaborate mess—costly, slow, and demanding of much maintenance. Coming into this new knowledge which proves so distressing—that our dearest companion, the automobile, is completely incompatible with our health and well-being—how long will we cling to it in this embrace of death?

We have noted that long after the exhaust smells have wafted away the hydrocarbons, in the presence of nitrogen oxides and that much publicized California sunlight (or anybody's less glorious, less publicized sunlight), keep on producing a veritable legion of lethal agents to inhabit the smog.

No American city is spared. Two years ago in San Francisco we were driven by burning eyes to flee the city to the skyline drive high above. From that height we looked down on a city swathed in a thick mustard-colored robe. This was a blanket manufactured by man himself, which he had drawn down over the once sparkling countryside and golden bay, and it was slowly smothering him to death. How long, we wondered, will man continue to sacrifice his cities, his enjoyment, his life, because of the fetid breath of a monster never built for meandering in city streets with their stops and goes and halting, jammed traffic? An ungainly, unadaptable monster whose 80 mile an hour cruising speed was geared to the long sweep of thruways and not to the bumper-bumper creep among the city's canyons?

If we would still cling to our 300-horsepower monsters in city lanes built to accommodate at most a 4-horsepower carriage, we have but two choices. Shall we choose mass poisoning in a slow fashion by the vast armies of hydrocarbon derivatives or a faster (perhaps more dashing) suicide from the nitrogen oxides? Apparently, technological difficulties so far obviate relief from both.

This small book, intent on presenting an overall view of our most pressing modern problems, does not pretend to medical authority regarding the effects of air pollution on the living community; many health authorities, specialists in the field, have performed this work with shocking competence. But one piece of medical research will be noted. As early as 1957 Dr. Paul Kotin, pathology professor of the University of Southern California, reported at Yale Medical School his findings of five years of study on laboratory animals. A group of mice exposed to the day-by-day Los Angeles air developed one and one-half times as many lung cancers as those who breathed clean air. Abnormal changes were found in the lungs of experimental animals after only a few months of this Los Angeles brew. Concluded the scientist: "Similar rapid changes can be expected in human beings, with many air pollutants combining to make the lungs more susceptible to cancer in a shorter time than previously believed."

Dr. Thomas P. Manusco, industrial hygiene chief for the Ohio Health Department, told the 1958 National Conference on Air Pollution that the urban lung cancer rate "increases by the size of the city." Since these reports, studies have confirmed the findings.

A man is not a mouse—usually. He has a bigger body; he may have more resistance, slower reaction. But every day the air a human breathes comes in direct contact with an area twenty-five times as great as his exposed skin area. This is the exposure surface of the tender membranes that line his lungs. Dr. Morris Cohn impressed on a New York Joint Legislative Committee on Air Pollution that while "man consumes less than ten pounds of water, fluids, and food daily, yet he requires over thirty pounds of air in the same period—and thirty pounds of air is a lot of air, 3,500 gallons of it!" It is not necessary to be an alarmist to conclude that man is not immune to what affects the mouse. Indeed, physiologically, he is more like the mouse than he wants to believe.

A cartoon of a few years ago presenting the inside of a U.S. weather station of the future depicted the weatherman making the following report: "Our latest analysis of the stratosphere, Dr. Figby! . . . 21 percent hydrogen, 7 percent oxygen . . . and 72 percent automobile exhaust fumes!" By 1985 this cartoon will no longer amuse. Predicts sanitation expert A. C. Stern of the Taft Sanitary Engineering Center, Cincinnati: By 1985 the U.S. Weather Bureau will be issuing daily air pollution reports as well as weather forecasts. People will be "more interested in whether it will be safe to breathe than whether it will be rainy or sunny." Our only comment here is that his prediction is for ten—probably fifteen—years too late.

One more word about our shiny master, the motorcar: As motors are stepped up for higher compression, year by year, nitrogen oxides are stepped up also. And as gasoline manufacturers vie for more "pick-up" by adding new substances like tetraethyl lead and nickel to the gasoline, these extremely toxic substances are also added to our atmosphere. The insane competition for speed and power bows neither to safety nor to health.

It is unreasonable to blame the manufacturers. In the end they put out what the public demands. Indeed, some of them are ahead of public demands as a matter of company pride. Some persons claim that the industry has the know-how to combust completely the fuel within the cylinder and does not do it. But how many American buyers, when dickering for a new car, ask anything more about an engine's performance than horsepower and the fast getaway—and perhaps mileage per gallon?

True, there is the gas-turbine engine that would get rid of all the nitrogen oxides and most of the smog. It will also burn practically anything. But will Americans accept (as the Russians are now doing) conversion to an engine that gives slow starts, noisier action, a trifle less "guts"? Detroit is convinced the customer prefers the fast jump to a long and happy life.

But the Man from Mars, standing beside a superhighway as the shiny monsters hurtle by at 75 or 85 miles an hour, the drivers bent tensely over their steering wheels with riveted, lusterless eyes that see nothing but the pavement ahead, wonders what it is all about. What is this strange, strained Earth creature getting out of life? It is evident that he is not enjoying himself. And the price is high.

Air pollution in America is so varied, so complex, so changing a problem that scientists have so far only begun to scratch the surface. It differs from city to city; turbulence, topography, sunlight, whether the majority of city dwellers burn coal or oil, the industries nearby, the number of automobiles and trucks—all these determine the quotient.

Typical of the great cities is the plight of Boston, described by Commissioner of Public Health Alfred L. Frechette. He estimates that 2,600 tons of solid contaminants are dumped daily over the central 100 square miles of the city. A basic problem is the open burning of automobile bodies. Yet an effective city incinerator that would reduce by 85 percent what goes into it costs about $5.00 a ton to operate, compared to the $1.50 a ton cost of open-air burning, or $3.00 a ton in a sanitary land fill operation. What shall the city fathers do?

There are two common denominators for air pollution from place to place: *first,* all air pollution is harmful, not only to people but to animals, plants, buildings, bridges, crops, and goods; and, *second,* all types are increasing everywhere at such a rate that researchers are left gasping. "Growing efforts to cope with this evil are having difficulty keeping pace with its fresh manifestations."

What, ask the worried officials at the Boston Museum of Fine Arts, can they do about the black spots which have appeared all over the 300 or 400 priceless and irreplaceable bronze artworks—some of them from the first century—spots on the green patina that come, presumably, from the sulphur oxides in the Boston air? The sulphur oxides evolve from the burning of coal and fuel oil, as well as from gasoline, and their effects cannot be reversed, say officials. Or maybe the blackening comes from ozone? Nobody is quite sure what pollutant causes it, but it is decimating the Museum's treasures.

And what about the worsening air pollution in the region of the booming steel mills? Advanced science now employs oxygen streams to hurry up metal "cooking," and these cause enormous clouds of reddish-brown, acrid smoke that shuts off sunlight and air, has an evil smell, and cloaks and stains homes nearby. An electrostatic precipitator to eliminate most of the smoke and smell costs $1,000,000 per furnace, and even with greatly stepped-up steel production (and, we would assume, greatly stepped-up profits) nobody wants to pay out a million dollars each for ten or twenty furnaces in a row.

Aerial spraying of crops not only poisons rivers but may drift on winds for many miles to lodge in people's lungs. Only recently DDT has been found as a pollutant in the air far from the agricultural lands where it was employed. Nuclear explosions added the pollutants of fallout to our atmosphere, radioactive isotopes that will still be descending on us in the rains for years to come.

In this connection the authors wrote in 1959 a small volume called *Our New Life with the Atom,* which raised several questions that scientists, then in the first blush of dazzling atomic predictions, were determinedly ignoring. One of these questions was: What would fallout—including radioactive iodine-131 from the multitude of Nevada tests—do to the bones and thyroids of children living nearby? The Atomic Energy Commission had been smugly assuring Americans and the entire world that the fallout it was causing was completely harmless. For example, an AEC spokesman on May 13, 1957, stated that the bomb tests would not have "the slightest possible effect" on humans.

On December 3, 1965, eight years later, an Associated Press release, originating in Salt Lake City, Utah, read as follows:

> *Atomic blasts in the 1950's are suspected of affecting thyroid glands in a group of Utah children soon to be given medical tests in Salt Lake City. There are nine children to be examined at the University of Utah Medical Center. Southwestern Utah, where the children live, is crossed by winds from southern Nevada. . . . The children are 10 to 18 years old. . . . All have nodules, or small lumps, on their thyroid glands.*

There is an ominous resemblance between the protestations of the "harmless fallout" experts of a few years ago and the "harmless insecticides" experts of today.

And what are we planning to do, wonders Dr. Columbus Iselin, Director of the Woods Hole Oceanographic Institution, about the possibly catastrophic effects of carbon dioxide on our weather? Modern technology is releasing great new volumes of this gas into the Earth's envelope, even while understanding that this gas is, over a period of time, a drastic climate changer. Man today is altering his environment faster than ever before, and little of it, to date, has been to the good.

Air pollution, like water pollution and all the other examples of our deteriorating conditions of living, is a by-product of too many people with too much push in the direction of economic progress and not enough in the direction of social progress. In a technology of surpassing wonders, wonders of dizzying impact, we still bear the body, psychological reactions, and evolutionary status of Cro-Magnon. Physically, we are eminently better adapted to cave living than to space living and will probably continue, if spared, in this retarded development for another million years.

Thus, in a civilization whose air may be composed of 21 percent natural gases and 79 percent automobile and other combustion gases, we shall be burdened with lungs that demand an outmoded mixture of some 79 percent plain nitrogen and 21 percent oxygen. We are trying to adapt a prehistoric physiology to an ultra-modern technology and losing on every front. Paleontology tells us that unless we switch and learn to adapt our technology to our prehistoric bodies, we shall perchance pay the price of the dinosaur and the other unadaptable life species which have preceded us.

The depth of our present ignorance concerning the hundreds of killer gases that we are generously releasing into our air blanket (and which, contrary to most belief, will not simply waft away out into space somewhere) remains the most incredible and awesome element of our entire way of life. Such poisons will linger in and densify our troposphere forever or until transformed into other substances. Their effects are therefore both acute and chronic.

Crash programs to learn more about air pollution and how to ameliorate it are in progress at all governmental levels. And in this field, as in others, scientists are beginning to desert the torn standard of combat against nature, to rally around the more solid standard of learning to work with her or at least to use her help.

For example, Dr. Chauncey D. Leake, Assistant Dean of the College of Medicine at Ohio State University, has called for extensive planting of trees and other green things ("Maybe 10 trees for every automobile and 100 for every truck") to depollute the air. In this proposal he is strongly supported by Dr. Philip L. Rusden of the Bartlett Tree Research Laboratories. While humans inhale oxygen and exhale carbon dioxide, trees take in carbon dioxide and discharge oxygen, greatly helping to purify the air.

The extensive sums of money put into research by industry (including the American Petroleum Institute) throughout the nation to end pollution both of the common waters and the air are not generally known or appreciated. Leonard A. Duval, President of Hess von Bulow, Incorporated, of Cleveland, is only one of the many industrial magnates with a firm dedication toward cleaning up pollution. "Whenever I see a cloud of ugly brownish smoke pouring from a steel-mill stack," says Mr. Duval, "or when I see a stream of discolored water pouring from a plant or factory into a creek or river, I squirm a little and say to myself, 'Look at all those dollars going to waste. I'm going to get some of them.' "

Already industry has learned how to recover many millions of dollars' worth of light oils, ammonia, and materials from coking coal —materials that go into drugs, plastics, chemicals, and many more products. Now Mr. Duval has dredged nearly 35,000 tons of exceptionally rich iron—washed-away mill scale—out of the Mahoning

River near Warren, Ohio, for which he obtained almost $400,000. He is erecting another dredge and four-story processing plant at the cost of about $300,000 at Niles, Ohio, to work another 4,000-foot stretch of the river for a possible additional $400,000. Says Mr. Duval: "There are millions lying on the bottoms of these mill-town rivers and creeks waiting for someone to pick it up."

In like manner the massive brownish clouds of smoke that belch from modern steelmaking furnaces contain iron and zinc sulphides that persistent research will teach how to reclaim. Even the "fly-ash" emitted by coal-burning electric plants might have a use. When air and water pollutants are made truly commercial—that is, when scientists establish a cash-redemption value for each—then the nation's waters and air will be cleaned up with alacrity.

But research, like art, is long, while time continues to be fleeting. Properly, the attack on our ignorance must be spearheaded by the national government. Not only would it be sheer waste for each community or state to duplicate each other's efforts; the call now is for such highly specialized atmospheric scientists, medics, chemists, engineers, meteorologists, etc., that smaller agencies of government could not easily recruit the talent called for in our emergency.

Once the facts are known and the solutions made available, the problem of applying those understandings is primarily local. As noted, one community may contend with a copper smelter, another with a chemical plant. One may suffer from atmospheric inversion demanding inflexible traffic limitations; another may have a soft coal problem. No distant official would be likely to work out the ingenious economical and effective program for trash burning, for instance, that Miami, Florida, did. There the incinerator will pay for itself by providing steam for a hospital and custom disposal for nearby Miami Beach.

There is an important role for the states, however. Enabling laws must be passed, rigid standards laid down, specialists provided for the smaller communities, and both intercommunity and interstate or regional problems attacked. The chief complaint of municipalities is that they have no "yardstick" by which to measure violations. It was partly to this end that an Air Pollution Control Board was created in New York State. And unless extensive education is carried on by all levels of government, there will not be the necessary public support to impel the large but needed outlays of money by either government or industry.

It took the Triangle Shirtwaist fire in 1911 to jar New York and then other states into enacting a labor law. It took a circus fire in Hartford to rouse us to more action. It took the near calamity of drought in the East to start pulling in the pollution violators for court hearings. We have already had some wholesale executions by air pollution in London, Los Angeles, and New York City on occasions of atmospheric inversion that hugged the smog close to the earth for a few days at a time. How big must the slaughter be to get real action?

Certainly what we need most is deglomeration of people. Excessive pollution of air as well as of water can deglomerate, even decimate, cities unhealthily crowded with men and cars. But there must be more pleasant ways.

One way less unpleasant than mass biocide by gas asphyxiation would be to lean a trifle harder on that phase of research seeking to produce a combustion engine that completely combusts its fuel and emits a minimum of poisons. Our present combustion engines waste oceans of fuel every year. Can the gas turbine engine be perfected? Chrysler's directional turbine nozzle is a smart advance in fuel efficiency and is hopeful. But why, ask the automobile industry's Members of the Board behind closed doors, throw in a wrench when you've got such a good thing going? No reason at all.

But the White House has given out a hint that a reason may be made. The President's Science Advisory Committee, newly alarmed about the leads and additives that have pushed air pollution to critically high levels, has been mumbling (not too indistinctly) about tighter federal controls under a sort of Food and Drug Administration type setup. They are going so far as to suggest that the time is coming "when it will be necessary to get rid of the present engine and fuels altogether." They have asked automakers seriously to "mull the idea of scrapping present engines and powering cars with non-toxic fuel cells instead."

This is no small request. Fuel cells, while they would completely do away with auto smog and all its train of miseries, are rather far from perfection; indeed, they are far from any practical application. The fact is, the condition of our cities' air is so bad we cannot afford to wait for them. What, then?

Since most families have two cars already it is suggested that one of them be a small electric cart for city driving and the other a high-powered machine for the road. No, the electric cart will not leap forward like a rocket at the green light, but consider this: No more poisonous vapors, odors, smoke clouds, corrosion, gluey oils, inflammatory gasoline; accidents cut to a tenth, less noise, easier parking, the innovation of a relaxed kind of driving (not to mention the cut in incidence of lung cancer, bronchitis, heart conditions, and smarting eyes).

There are, in all, from 8,000 to 10,000 tons of gases, vapors, and solids being thrown into a large city's air every day—a generous two-thirds of it from the automobile—to saturate the lungs of roughly two-thirds of the nation's population. Years ago former Surgeon General Leroy T. Burney declared categorically that there is a "definite association between community air pollution and high mortality rates," a fact that is today universally accepted.

While cars get faster and longer, lives get slower and shorter. While Chrysler competes with Buick for the getaway, cancer competes with emphysema for the layaway. This generation is indeed going to have to choose between humans and the automobile. Perhaps most families have too many of both.

James Ramsey
WILDERNESS AND CONSERVATION

The Sierra Club was founded in 1892 to pursue a twofold purpose that is still central to the organization: (1) "exploring, enjoying and rendering accessible the mountain regions of the Pacific Coast," and (2) "to enlist the support of the people and government in preserving the forests and other features of the Sierra Nevada Mountains." Although still headquartered in San Francisco, the Club has extended its enjoying and preserving to the entire nation. John Muir was its first president and held office until his death in 1914. The Club has been instrumental in landmark wilderness preserva-

From "Wonderland Revisited," by James Ramsey, in the *Sierra Club Bulletin,* Vol. 54, No. 10 (October–November 1969), pp. 10–13. Reprinted by permission.

tion actions, including the establishment of Yosemite National Park, the 1916 National Park Service Act, the preservation of Echo Park in 1950, and the National Wilderness Preservation System in 1964. The Club's greatest defeats have been in the loss of Hetch Hetchy Valley to a reservoir and Utah's Glen Canyon to Lake Powell in the 1960's. The Club gained national recognition in recent years because of the militant watchdog leadership of David R. Brower, now of the very aggressive Friends of the Earth organization. Along with its outspoken and respected journal, the Sierra Club Bulletin, the Club has sponsored major Biennial Wilderness Conferences and regularly publishes splendid books about the wilderness, untouched or damaged. The following selection was written by James Ramsey, editor of the Bulletin.

"Would you tell me, please, which way I ought to go from here?" asked Alice.
"That depends a good deal on where you want to get to," said the Cat.
"I don't much care where—" said Alice.
"Then it doesn't matter which way you go," said the Cat.
—From Alice's Adventures in Wonderland

Move over Alice, you've got a couple of hundred million Americans for company. We don't know where we are going or where we want to get to either. Like the guests at the Mad Tea-Party, we eat and move on around the table to the next place-setting, leaving the dirty dishes behind, never once facing-up to the question of what happens when we complete the circle. Our national priorities are caricatures without substance, more illusory than the Cheshire Cat. We plan ahead with the same appreciation of the future as that enjoyed by the Mad Hatter. It is always six o'clock.

Consider these scenes from the theater of the absurd:

—At its National Reactor Test Station in Idaho, the Atomic Energy Commission is burying radioactive waste that will be deadly to any living thing exposed to it for the next 1500 years. The burial ground is in the same general vicinity where three rivers flow out of the mountains and disappear into the desert floor.

—This year, the U.S. Army decided it was overstocked with biological warfare weapons and as an economy measure proposed to dump a quantity of them into the Atlantic Ocean. While Army spokesmen argued the merits of this macabre plan, an entire train-load of the stuff was parked directly off the end of the main Denver airport runway.

In the rush to make it we have created a wonderland of rhetoric, rationalization and double-think that not even Lewis Carroll could have imagined.

—During the Redwood National Park controversy the logging interests complained bitterly that the establishment of a tiny parcel of land as a park would interfere with their sustained yield program. At the same time, they were going ahead with plans to log the last commercial old-growth redwoods on earth within the decade, knowing full well that the growing period is too long, the supply too short for a sustained yield program.

—The dam builders were stopped from flooding the Grand Canyon by an aroused public, and in response proposed to plug Hell's Canyon, the deepest gorge on the North American continent.

—As a deadly twilight descends over our cities, highway departments across the land mid-wife the birth of new freeways, which spawn more automobiles, which demand more freeways . . .

—And the President of the United States, faced with the greatest domestic crisis his or any other administration has ever had to deal with, talks of national pride and pushes the development of a monstrous supersonic passenger jet that is neither needed nor wanted by his constituency.

These bizarre actions and hundreds more like them are accompanied by a chorus of Orwellian Newspeak aimed at convincing all of us of the ultimate truth—the stink of a pulp mill is actually the perfume of progress. Humpty Dumpty couldn't have put it better.

The latest symptom of this growing national malaise is taking place now in the mad scramble to explore and exploit the oil reserves of the Alaskan Arctic. The discovery of oil at Prudhoe Bay has been well publicized, as has the near-billion-dollar sale of leases by the State of Alaska in September. Under this barrage of good tidings the plea for ecological sanity is but a squeak. The race to nowhere is on. The long distance runners of this American Dream set piece are the oilmen, the rugged sourdoughs of the late twentieth century. Armed with a bank of computers, a fleet of Lear jets, engineering degrees and a 27.5 percent oil depletion allowance, a gift from a grateful government, they bravely face the perils of the Arctic. Theirs is a mission of Urgency and Importance. They speak darkly of troubles in the mid-east, of the need for a new domestic oil supply, diminishing national reserves, a viable economy, and the spectre of war.

They do not speak of the question of why we should pump billions of barrels of future air pollutants from an area that will almost certainly be irreparably scarred in the process; an area that if left untouched would be a far more valuable national resource in future years than all the oil beneath its surface. But not now, and oilmen have not been programmed to think beyond now.

Wonderland permits no such heresy. The rhetoric says Alaskan wilderness is infinite and therefore indestructible. The rhetoric says there is no connection between the extractors of a resource and its ultimate use, and anyone suggesting otherwise is demented and very possibly un-American. But rhetoric and reality do not agree. Given the pressures of population and the present state of our technological juggernaut, a relatively pristine Alaska should last at most about twenty years. Coincidentally, this is also about the length of time that urban air will still be breathable, unless something is done, and done quickly.

But the oilmen will not plug up their Alaska wells and go home. Their logic of exploitation has never progressed beyond the level of reasoning expressed by Mallory when asked why he wanted to climb Mt. Everest. "Because it is there" is a fitting reason for climbing a mountain, but shortsighted in the extreme as a basis for establishing a huge industrial operation on the last wild, untouched frontier of this continent, a frontier as fragile as it is beautiful.

It may seem inaccurate to describe the Arctic tundra as fragile with all the powerful environmental forces at work there. But it is precisely these forces—the eternal cold, the long winter nights, the howling blizzards, and the permanently frozen subsoil—that make the Arctic ecologically and esthetically fragile. The complex food chains of the temperate zones are reduced to the bare minimum in the Arctic; often an entire food web consists of only a few species of plants, plant eaters, meat eaters and scavengers. An iron law of nature is that variety means survival, and the Arctic is short on this kind of variety.

Left to themselves, the ecosystems of the Arctic tundra function perfectly, and have done so for millions of years. But they do not have the capacity to withstand any but the most careful incursions of man, particularly mechanized man. The passage of a single, tracked vehicle over thawed tundra may leave a scar that will last for decades, or forever. No one really knows what effect heavy concentrations of humans and their machines will have on the migratory habits of Arctic wildlife such as the caribou, or on already endangered species that live there. In fact, the single most critical issue of the whole Arctic exploitation question is the fact that so little is known about Arctic ecology at this time that it is impossible to assess potential damage.

Given this incontrovertable fact, and in the absence of demonstrable proof that there is any urgent national need for Arctic oil in the immediate future, isn't it reasonable to conclude that the extraction of oil, if it is done at all, should proceed only after careful and extensive ecological studies are made? No, it is not; not in Wonderland.

> *"But I don't want to go among mad people," Alice remarked.*
> *"Oh, you can't help that," said the Cat: "We're all mad here. I'm mad. You're mad."*

The oil companies, the State of Alaska, and the Federal Government in the person of the Secretary of Interior, are all hell-bent to begin pumping oil out of the ground as soon as possible. They give lip service to the need for ecological studies, but if present plans are consummated, all but cursory investigations will be after the fact, not before.

The oilmen are full of brobdingnagian schemes to transport the oil out of the roadless Arctic to the prime markets in the lower forty-eight. They are predictably products of the engineering mentality so prevalent in the extraction industries where biological considerations are always placed at the bottom of the priority list. One plan calls for a fleet of super tankers equipped with special ice-breaker bows which would transport the oil directly from Prudhoe Bay to the major east coast markets. A "successful" test was conducted this summer with the super-tanker Manhattan reaching the Arctic area, but only after being unstuck from the ice on several occasions by Canadian ice-breakers. No tests were made, of course, on what would happen to Arctic marine life, including polar bears and seals, if one of these oil-bloated ships should duplicate the Torrey Canyon disaster.

Another proposed method of transporting oil to the states is an environmental horror that would put a huge pipeline across the state —an 800-mile long scar from Prudhoe Bay to the port of Valdez. Financed by a consortium of oil companies—Humble, Atlantic-Richfield and British Petroleum—the four-foot diameter pipeline would carry 1 million barrels per day and require a massive construction project involving roads, pumping stations and the pipeline itself cutting directly across the Arctic slope and through the incomparably beautiful Brooks Range. In spite of some serious doubt about the pipeline's technical and economic feasibility the oilmen are already unloading sections of pipe at Valdez.

The oilmen have asked Secretary of Interior Hickel to unfreeze right-of-way areas on Federally owned land presently held up pending settlement of Native Land Claims. The Secretary has displayed a remarkable eagerness to comply with their request. His department has performed one of the miracles of our time by preparing a list of stipulations governing the construction and operation of the pipeline based on ecological information that isn't even known yet. The stipulations and a request for approval to unfreeze the right-of-way were sent to the Senate and House Interior Committees on October 1.

As quoted in the *Oil Daily,* the Secretary embellished his request with a statement that reached new heights in jabberwockian splendor. "The stipulations," he said, "will insure that the wildlife and ecology of the Arctic, along with the culture and opportunities of Alaska's native citizens, will be enhanced." What, on the little that remains of God's green earth, does he mean by that? Federally funded Medicare for Caribou? How can an 800-mile-long mechanical monstrosity "enhance" the ecology of the Arctic? And as for improving the culture and opportunities of Alaska's native citizens, if the Secretary means by *their* standards, not ours, it will be the first time in this nation's history that any of its aboriginal inhabitants have been introduced to such a novel concept.

The Secretary also noted that the stipulations "are designed to meet *all* of the environmental and ecological goals set forth by the department, based on research by its own scientists, independent authorities and public hearings held in Alaska." Nonsense! If the pipeline is allowed at this time it will be installed on a trial and error basis, pure and simple. Most of the studies on just the mechanical phases of pipeline operation in the Arctic—investigations into the effect of the pipeline on permafrost and vice-versa, for example— have not been completed. Hardly any biological studies have even been started. Throughout the stipulations the emphasis is placed on remedial action *after* the pipeline ruptures or malfunctions, not on how to prevent it from happening in the first place.

The pipeline fiasco would be ludicrous if it were not the precursor of what is to follow in Alaska. If indeed, Alaska is allowed to be exploited with the mindless planning and narrow economic justifications that have characterized so-called progress in America, the last chance to bring some sanity to our choice of national priorities will have been lost. Alaska is a pivotal area, in a pivotal time. The battles that will be fought in Alaska will not be just to save a chunk of land or a specie of wildlife—they will be to decide what things are really important for human beings to continue living on this planet. If these decisions are left to those who have made them in the past, it will not be just another ecological battle lost; it will be, in the real meaning of the term, inhuman.

"First, the Dodo marked out a race-course, in a sort of a circle, and then all the party were placed along the course, here and there. There was no 'one, two, three and away,' but they began running when they liked, and left off when they liked, so that it was not easy to know when the race was over. However, when they had been running a half-hour or so, the Dodo suddenly called out 'the race is over' and they all crowded around it, panting and asking, 'But who has won?' "

IV "ON A CLEAR DAY YOU CAN SEE FOREVER"— ALTERNATIVE SOLUTIONS

Paul B. Sears
A NEW ULTIMATE SCIENCE?

Paul B. Sears is a botanist and conservationist at Yale. He has become a spokesman for biological scientists who believe that their discipline has wrongly been neglected in the modern quest for new centers of value. Western man, they argue, has improperly downgraded man's organic physical nature in his quest for "spiritual" ideals. This is a false dualism which has intensified and accelerated the environmental crisis by its neglect of fundamental and inescapable biological facts. In 1966 Dr. Sears published The Living Landscape, *in which he called for the recovery of links between man and nature in order to save humanity from enclosing itself in a sterile and mechanistic environment. Properly understood, the life chains, energy systems, and other forms of human dependency upon the environment are rich with ideas of ethical obligations and relations. The following selection first appeared in the journal* BioScience *in 1964. Dr. Sears called ecology "a subversive science" because it seemed to upset Western man's habitual refusal to find meaning in his biological condition.*

My choice of title ["Ecology—A Subversive Subject"] is not facetious. I wish to explore a question of growing concern. Is ecology a phase of science of limited interest and utility? Or, if taken seriously as an instrument for the long-run welfare of mankind, would it endanger the assumptions and practices accepted by modern societies, whatever their doctrinal commitments? To this end, I propose to consider its position in the world of science, the field of education, and the arena of practical life.

Disregarding for a moment the judgments of ecologists themselves, I have encountered two criticisms. The first is that ecology is merely a matter of emphasizing the obvious. It merely assembles matters of common knowledge and attempts to invest them with status by means of a special and not very winning vocabulary. The second represents an opposite extreme. Conceding the ultimate importance of understanding the great pattern of life and environment, sound logic and practice dictate that we must *first* get on with the infinitely detailed analysis of the many factors involved. In other

From "Ecology—A Subversive Subject," by Paul B. Sears, in *BioScience*, Vol. 14, No. 7 (July 1964), pp. 11–13. Reprinted by permission. Notes to the original have been omitted.

words, we do not yet know enough about the bricks and mortar to get on with the building. I am certain these attitudes have been reflected in the appointment programs of influential schools.

*   *   *

. . . The present position of ecology in education is an eloquent fact. I have lectured on more than six-score campuses, generally under circumstances that permitted a look around, thanks to the hospitality encountered. Almost without exception, I have found audiences of students, faculty, and citizens responsive to discussion of the nature and significance of ecology. Yet, with too few exceptions, I have found the instructional juggernaut creaking along in conventional paths so far as this subject is concerned.

The several introductory science courses which should here if anywhere have a common intellectual bond continue to go their separate, insulated ways, each with a reverent eye upon the minority—its future "majors." Incidentally, but by no means trivially, I know of cases where potential recruits of high scientific promise have been lost by this frozen practice. Even the millions spent on course improvement, fruitful as they may have been, have done more to intensify teaching of the individual sciences than to integrate them.

It is also my observation that the majority of biological curricula have been unable to wriggle free from the trap of convention. In the typical introductory course, ecology tends to come in as a chapter along the way rather than as a unifying philosophical point of view. With few exceptions, the builders of textbooks follow this pattern, thereby perpetuating it. Since their publishers must keep a shrewd eye on the buyer's market, it is hard to assess final responsibility.

*   *   *

Now training and education form a continuum whose charter derives from the needs of the culture. The group must survive under satisfying conditions and, to that end, so must the individual. Thus the elements of necessity and satisfaction, of advantage and delight, are intertwined. As in a crude handmade rope, one strand may at times conceal or suppress the others. Military and civic necessity led the Romans to foster the military arts, lucid communication, engineering, and medicine. Aristocratic delight was a large element in the brilliant science of 18th century France and lingered on to inspire Darwin.

What we call higher education is compounded of all these factors. It reenforces the training needed for social and individual advantage by intellectual challenge or respectability. Does ecology qualify under these latter rubrics? Genetics, its twin, certainly has done so. To answer this question fairly, we must lay aside any arrogance with respect to the simple pioneer work of the early ecologists, or even its relevance to certain steps in the present learning process. Astronomers do not laugh off Kepler or Copernicus or even Ptolemy, and, if chemists were to speak too lightly of the Greek elements—earth, air, fire, and water, their students would miss the very basis of the analytical process.

*　　*　　*

. . . Ecology, despite its fragmentary progress beginning with the environmental relations of plant life, is a study of the entire ecosystem. Of this system, man is not just an observer and irresponsible exploiter but an integral part, now the world's dominant organism. He has come into the system and survived thus far by the bounty of that system plus his own marvelous power of adjustment. Even so, the historical record is replete with his failures.

By its very nature, ecology affords a continuing critique of man's operations within the ecosystem. The applications of other sciences are particulate, specialized, based on the solution of individual problems with little if any attention to side effects and practically uncontrolled by any thought of the larger whole.

The problems to which scientific technology has addressed itself are real enough. No farm boy who has forked manure, milked many cows by hand, or trudged a long distance to school would question this. No parent who has seen his family suffer the ills of overcrowding, poverty, and hunger would question it—and neither would the producer who has been able to substitute power machinery for slow and laborious hand labor, nor the householder who recalls the days before telephones and modern plumbing. The handicaps of distance, disease, hard muscular labor, and insufficient food cannot be ignored.

But technical improvements on the farm have driven millions off the land and into the city where the progress of invention has made the profitable use of manpower less and less necessary. Within the cities themselves, the penalties of overcrowding daily become more obvious as the maintenance of order, education, and health become increasingly difficult. Mass production, with its steadily increasing drains upon energy and materials, is now so efficient that industry has to exert itself to create demand, while government is enmeshed in legerdemain to increase purchasing power through devices that were old when kings diluted the gold in their coins. Meanwhile our efficient industry and domestic sanitation push their wastes into air and water—painful mockery of the quiet recycling of materials in the ancient pattern of nature.

I know of no more eloquent presentation of the situation than that by the great humanitarian Dr. Alan Gregg shortly before his death. After a lifetime spent in bringing the benefits of modern hygiene and medicine to the edges of the habitable planet, thereby mightily helping to reduce the death rate and prolong life, he reviewed the resulting ecological imbalance with grim dismay. Drawing upon his professional knowledge, he could only compare the uncontrolled increase in human numbers and the resulting disorder with the spread of cancer cells within an organism.

To me, at least, it is disturbing to hear the current glib emphasis on economic "growth" as the solution of all ills. Growth, in all biological experience, is a determinate process. Out of control, say by pituitary imbalance, it becomes pathological gigantism and by no means the same thing as health. With the concept of a healthy economy there can be no quarrel, but to equate this with an ever-expanding, ever-rising spiral is to relapse into the folly of perpetual motion, long since discredited by a sane understanding of energetics.

I have no wish to continue a depressing recital, even when I recall that the value of stored farm surplus is exceeded by that of military materiel designed and manufactured since the end of World War II but already obsolete. Honorable men are now wondering whether it has been ethical to introduce modern death-control into areas where there is no control of human numbers. I have not yet heard them justify the continuance of suspicion and the multiplication of armaments with the double objective of keeping industry prosperous and eventually reducing global population, but this may be a short step. I hope not.

One could quarrel with the profit motive or political philosophy as a source of present troubles, but the trouble lies deeper. The individual must see a reasonable advantage to himself to do his best. The leader of world Communism saw the great ears of Iowa corn but not the results of exploiting our dry steppes, and so he needlessly repeated that folly. The British Labor party, when in power, spent treasure on the ground-nut adventure when any ecologist worth his salt (and Britain has some of the best) knew it would not work. Likewise he could have warned the industries and housing developments that suffered great damage during the New England flood of 1955 of the hazards of their disastrous choice of sites.

My fellow ecologists are insistent on the need for more study of ecosystems, and they are correct. The planet is so vast and varied that our knowledge of it is quite imperfect. Blanket solutions from distant centers of power can seldom be trusted. It was this fact, combined with my belief that, as a matter of political health, the citizen must face the facts where he lives, that led me in 1935 to suggest the need for ecologists at the local level. I have lately been heartened to hear this suggestion rise anew from others. Let me repeat that the benefits of applied ecology can be realized only through widespread general understanding rather than through ordinary commercial channels. Closely related to this are, in my judgment, some fundamental reforms in the educational process. . . .

Withal I am mindful of the complaint of a Congressman who said that every scientist who appeared before his committee believes that his cat is blackest. Viewing, as I do, science as it should be, a seamless fabric, this is not my point. Granting the need for continuing advances along the entire front, I am trying to say that we are not using vital knowledge we now have. And, because this kind of knowledge affords perspective to a peculiar degree, we cannot safely continue to neglect it, however much it may threaten some of our cherished practices in the interest of our long-time welfare.

Postscript

Lest my estimate of the impact which the rules of ecological experience might have on our present ideas and practices seem ill-judged, I submit the following questions:

What conclusion would you draw if you observed a population curve, similar to that of man, in any other organism?

What are the effects upon the ecosystem where profit to the developer is the sole check upon urban expansion?

What are the patterns of material and energy flow consequent upon modern technology?

What is known of the cumulative effects on human beings of the wastes and byproducts of our present culture?

What is known of the long-range effects of monoculture and heavy machinery upon fertile agricultural land?

What will be the consequences of locating highways and other permanent public works on strictly engineering considerations, without regard to long-range use potential?

Loren Eiseley

THE NATURALIST'S VISION

Criticism by biologists of the nature and destiny of the American environment has also led them to propose a new philosophy of existence, civilization, and values. Loren Eiseley's writings express a world view where man's biological nature is more intimately cross-indexed with man's ideals. Biologist and anthropologist at the University of Pennsylvania, Dr. Eiseley argues that life is still deeply governed by forces that have their roots in the sky, soil, and water around man. The contemporary task is to bring modern man to recognize this reality even while the traditions and habits of Western society lead him away from these biological foundations. His views have received widespread attention among both specialists and general readers through several eloquent books, including The Immense Journey *in 1957 and* The Firmament of Time *in 1960; the following selection is excerpted from the latter book. Eiseley describes man's biological heritage as a definitive, fruitful, and neglected center of meaning and value.*

If we examine the living universe around us which was before man

From *The Firmament of Time* by Loren Eiseley. Copyright © 1960 by Loren Eiseley. Copyright © 1960 by the Trustees of the University of Pennsylvania. Reprinted by permission of Atheneum Publishers.

and may be after him, we find two ways in which that universe, with its inhabitants, differs from the world of man: first, it is essentially a stable universe; second, its inhabitants are intensely concentrated upon their environment. They respond to it totally, and without it, or rather when they relax from it, they merely sleep. They reflect their environment but they do not alter it. In Browning's words, "It has them, not they it."

Life, as manifested through its instincts, demands a security guarantee from nature that is largely forthcoming. All the release mechanisms, the instinctive shorthand methods by which nature provides for organisms too simple to comprehend their environment, are based upon this guarantee. The inorganic world could, and does, exist in a kind of chaos, but before life can peep forth, even as a flower, or a stick insect, or a beetle, it has to have some kind of unofficial assurance of nature's stability, just as we have read that stability of forces in the ripples impressed in stone, or the rain marks on a long-vanished beach, or the unchanging laws of light in the eye of a four-hundred-million-year-old trilobite.

The nineteenth century was amazed when it discovered these things, but wasps and migratory birds were not. They had an old contract, an old promise, never broken till man began to interfere with things, that nature, in degree, is steadfast and continuous. Her laws do not deviate, nor the seasons come and go too violently. There is change, but throughout the past life alters with the slow pace of geological epochs. Calcium, iron, phosphorus, could exist in the jumbled world of the inorganic without the certainties that life demands. Taken up into a living system, however, *being* that system, they must, in a sense, have knowledge of the future. Tomorrow's rain may be important, or tomorrow's wind or sun. Life, in contrast to the inorganic, is historic in a new way. It reflects the past, but must also expect something of the future. It has nature's promise—a guarantee that has not been broken in three billion years—that the universe has this queer rationality and "expectedness" about it. "Whatever interrupts the even flow and luxurious monotony of organic life," wrote Santayana, "is odious to the primeval animal."

This is a true observation, because on the more simple levels of life, monotony is a necessity for survival. The life in pond and thicket is not equipped for the storms that shake the human world. Its small domain is frequently confined to a splinter of sunlight, or the hole under a root. What life does under such circumstances, how it meets the precarious future (for even here the future can be precarious), is written into its substance by the obscure mechanisms of nature. The snail recoils into his house, the dissembling caterpillar who does not know he dissembles, thrusts stiffly, like a budding twig, from his branch. The enemy is known, the contingency prepared for. But still the dreaming comes from below, from somewhere in the molecular substance. It is as if nature in a thousand forms played games against herself, but the games were each one known, the rules ancient and observed.

It is with the coming of man that a vast hole seems to open in nature, a vast black whirlpool spinning faster and faster, consuming flesh, stones, soil, minerals, sucking down the lightning, wrenching power from the atom, until the ancient sounds of nature are drowned in the cacophony of something which is no longer nature, something instead which is loose and knocking at the world's heart, something demonic and no longer planned—escaped, it may be—spewed out of nature, contending in a final giant's game against its master.

Yet the coming of man was quiet enough. Even after he arrived, even after his strange retarded youth had given him the brain which opened up to him the dimensions of time and space, he walked softly. If, as was true, he had sloughed instinct away for a new interior world of consciousness, he did something which at the same time revealed his continued need for the stability which had preserved his ancestors. Scarcely had he stepped across the border of the old instinctive world when he began to create the world of custom. He was using reason, his new attribute, to remake, in another fashion, a substitute for the lost instinctive world of nature. He was, in fact, creating another nature, a new source of stability for his conflicting erratic reason. Custom became fixed: order, the new order imposed by cultural discipline, became the "nature" of human society. Custom directed the vagaries of the will. Among the fixed institutional bonds of society man found once more the security of the animal. He moved in a patient renewed orbit with the seasons. His life was directed, the gods had ordained it so. In some parts of

the world this long twilight, half in and half out of nature, has per-
sisted into the present. Viewed over a wide domain of history this
cultural edifice, though somewhat less stable than the natural world,
has yet appeared a fair substitute—a structure, like nature, reasonably secure. But the security in the end was to prove an illusion.
It was in the West that the whirlpool began to spin. Ironically, it
began in the search for the earthly Paradise.

The medieval world was limited in time. It was a stage upon
which the great drama of the human Fall and Redemption was
being played out. Since the position in time of the medieval culture
fell late in this drama, man's gaze was not centered scientifically
upon the events of an earth destined soon to vanish. The ranks of
society, even objects themselves, were Platonic reflections from
eternity. They were as unalterable as the divine Empyrean world
behind them. Life was directed and fixed from above. So far as the
Christian world of the West was concerned, man was locked in an
unchanging social structure well nigh as firm as nature. The earth
was the center of divine attention. The ingenuity of intellectual men
was turned almost exclusively upon theological problems.

As the medieval culture began to wane toward its close, men
turned their curiosity upon the world around them. The era of the
great voyages, of the breaking through barriers, had begun. Indeed,
there is evidence that among the motivations of those same voyagers,
dreams of the recovery of the earthly Paradise were legion. The
legendary Garden of Eden was thought to be still in existence.
There were stories that in this or that far land, behind cloud banks
or over mountains, the abandoned Garden still survived. There were
speculations that through one of those four great rivers which were
supposed to flow from the Garden, the way back might still be
found. Perhaps the angel with the sword might still be waiting at
the weed-grown gateway, warning men away; nevertheless, the idea
of that haven lingered wistfully in the minds of captains in whom the
beliefs of the Middle Ages had not quite perished.

There was, however, another, a more symbolic road into the
Garden. It was first glimpsed and the way to its discovery charted
by Francis Bacon. With that act, though he did not intend it to be so,
the philosopher had opened the doorway of the modern world. The
paradise he sought, the dreams he dreamed, are now intermingled

with the countdown on the latest model of the ICBM, or the radio-active cloud drifting downwind from a megaton explosion. Three centuries earlier, however, science had been Lord Bacon's road to the earthly Paradise. "Surely," he wrote in the *Novum Organum,* "it would be disgraceful if, while the regions of the material globe, that is, of the earth, of the sea, and of the stars—have been in our times laid widely open and revealed, the intellectual globe should remain shut up within the narrow limits of the old discoveries."

Instead, Bacon chafed for another world than that of the restless voyagers. "I am now therefore to speak touching Hope," he rallied his audience, who believed, many of them, in a declining and decay-ing world. Much, if not all, that man lost in his ejection from the earthly Paradise might, Bacon thought, be regained by application, so long as the human intellect remained unimpaired. "Trial should be made," he contends in one famous passage, "whether the com-merce between the mind of men and the nature of things . . . might by any means be restored to its perfect and original condition, or if that may not be, yet reduced to a better condition than that in which it now is." To the task of raising up the new science he devoted himself as the bell ringer who "called the wits together."

Bacon was not blind to the dangers in his new philosophy. "Through the premature hurry of the understanding," he cautioned, "great dangers may be apprehended . . . against which we ought even now to prepare." Out of the same fountain, he saw clearly, could pour the instruments of beneficence or death.

Bacon's warning went unheeded. The struggle between those forces he envisaged continues into the modern world. We have now reached the point where we must look deep into the whirlpool of the modern age. Whirlpool or flight, as Max Picard has called it, it is all one. The stability of nature on the planet—that old and simple promise to the living, which is written in every sedimentary rock—is threatened by nature's own product, man.

Not long ago a young man—I hope not a forerunner of the coming race on the planet—remarked to me with the colossal insensitivity of the new asphalt animal, "Why can't we just eventually kill off everything and live here by ourselves with more room? We'll be able to synthesize food pretty soon." It was his solution to the problem of overpopulation.

I had no response to make, for I saw suddenly that this man was in the world of the flight. For him there was no eternal, nature did not exist save as something to be crushed, and that second order of stability, the cultural world, was, for him, also ceasing to exist. If he meant what he said, pity had vanished, life was not sacred, and custom was a purely useless impediment from the past. There floated into my mind the penetrating statement of a modern critic and novelist, Wright Morris. "It is not fear of the bomb that paralyzes us," he writes, "not fear that man has no future. Rather, it is the *nature* of the future, not its extinction, that produces such foreboding in the artist. It is a numbing apprehension that such future as man has may dispense with art, with man as we now know him, and such as art has made him. The survival of men who are strangers to the nature of this conception is a more appalling thought than the extinction of the species."

There before me stood the new race in embryo. It was I who fled. There was no means of communication sufficient to call across the roaring cataract that lay between us, and down which this youth was already figuratively passing toward some doom I did not wish to see. Man's second rock of certitude, his cultural world, that had gotten him out of bed in the morning for many thousand years, that had taught him manners, how to love, and to see beauty, and how, when the time came, to die—this cultural world was now dissolving even as it grew. The roar of jet aircraft, the ugly ostentation of badly designed automobiles, the clatter of the supermarkets could not lend stability nor reality to the world we face.

Before us is Bacon's road to Paradise after three hundred years. In the medieval world, man had felt God both as exterior lord above the stars, and as immanent in the human heart. He was both outside and within, the true hound of Heaven. All this alters as we enter modern times. Bacon's world to explore opens to infinity, but it is the world of the outside. Man's whole attention is shifted outward. Even if he looks within, it is largely with the eye of science, to examine what can be learned of the personality, or what excuses for its behavior can be found in the darker, ill-lit caverns of the brain.

The western scientific achievement, great though it is, has not concerned itself enough with the creation of better human beings, nor with self-discipline. It has concentrated instead upon things, and assumed that the good life would follow. Therefore it hungers for infinity. Outward in that infinity lies the Garden the sixteenth-century voyagers did not find. We no longer call it the Garden. We are sophisticated men. We call it, vaguely, "progress," because that word in itself implies the endless movement of pursuit. We have abandoned the past without realizing that without the past the pursued future has no meaning, that it leads, as Morris has anticipated, to the world of artless, dehumanized man.

*   *   *

"The special value of science," a perceptive philosopher once wrote, "lies not in what it makes of the world, but in what it makes of the knower." Some years ago, while camping in a vast eroded area in the West, I came upon one of those unlikely sights which illuminate such truths.

I suppose that nothing living had moved among those great stones for centuries. They lay toppled against each other like fallen dolmens. The huge stones were beasts, I used to think, of a kind man ordinarily lived too fast to understand. They seemed inanimate because the tempo of the life in them was slow. They lived ages in one place and moved only when man was not looking. Sometimes at night I would hear a low rumble as one drew itself into a new position and subsided again. Sometimes I found their tracks ground deeply into the hillsides.

It was with considerable surprise that while traversing this barren valley I came, one afternoon, upon what I can only describe as a very remarkable sight. Some distance away, so far that for a little space I could make nothing of the spectacle, my eyes were attracted by a dun-colored object about the size of a football, which periodically bounded up from the desert floor. Wonderingly, I drew closer and observed that something ropelike which glittered in the sun appeared to be dangling from the ball-shaped object. Whatever the object was, it appeared to be bouncing faster and more desperately as I approached. My surroundings were such that this hysterical dance of what at first glance appeared to be a common stone was quite unnerving, as though suddenly all the natural objects in the valley were about to break into a jig. Going closer, I penetrated the mystery.

The sun was sparkling on the scales of a huge black snake which was partially looped about the body of a hen pheasant. Desperately the bird tried to rise, and just as desperately the big snake coiled and clung, though each time the bird, falling several feet, was pounding the snake's body in the gravel. I gazed at the scene in astonishment. Here in this silent waste, like an emanation from nowhere, two bitter and desperate vapors, two little whirlwinds of contending energy, were beating each other to death because their plans—something, I suspected, about whether a clutch of eggs was to turn into a thing with wings or scales—this problem, I say, of the onrushing nonexistent future, had catapulted serpent against bird.

The bird was too big for the snake to have had it in mind as prey. Most probably, he had been intent on stealing the pheasant's eggs and had been set upon and pecked. Somehow in the ensuing scuffle he had flung a loop over the bird's back and partially blocked her wings. She could not take off, and the snake would not let go. The snake was taking a heavy battering among the stones, but the high-speed metabolism and tremendous flight exertion of the mother bird were rapidly exhausting her. I stood a moment and saw the bloodshot glaze deepen in her eyes. I suppose I could have waited there to see what would happen when she could not fly; I suppose it might have been worth scientifically recording. But I could not stand that ceaseless, bloody pounding in the gravel. I thought of the eggs somewhere about, and whether they were to elongate and writhe into an armor of scales, or eventually to go whistling into the wind with their wild mother.

So I, the mammal, in my way supple, and less bound by instinct, arbitrated the matter. I unwound the serpent from the bird and let him hiss and wrap his battered coils around my arm. The bird, her wings flung out, rocked on her legs and gasped repeatedly. I moved away in order not to drive her further from her nest. Thus the serpent and I, two terrible and feared beings, passed quickly out of view.

Over the next ridge, where he could do no more damage, I let the snake, whose anger had subsided, slowly uncoil and slither from my arm. He flowed away into a little patch of bunch grass—aloof, forgetting, unaware of the journey he had made upon my wrist, which throbbed from his expert constriction. The bird had contended for birds against the oncoming future; the serpent writhing into the bunch grass had contended just as desperately for serpents. And I, the apparition in that valley—for what had I contended?—I who contained the serpent and the bird and who read the past long written in their bodies.

Slowly, as I sauntered dwarfed among overhanging pinnacles, as the great slabs which were the visible remnants of past ages laid their enormous shadows rhythmically as life and death across my face, the answer came to me. Man could contain more than himself. Among these many appearances that flew, or swam in the waters, or wavered momentarily into being, man alone possessed that unique ability.

The Renaissance thinkers were right when they said that man, the Microcosm, contains the Macrocosm. I had touched the lives of creatures other than myself and had seen their shapes waver and blow like smoke through the corridors of time. I had watched, with sudden concentrated attention, myself, this brain, unrolling from the seed like a genie from a bottle, and casting my eyes forward, I had seen it vanish again into the formless alchemies of the earth.

For what then had I contended, weighing the serpent with the bird in that wild valley? I had struggled, I am now convinced, for a greater, more comprehensive version of myself.

I am a man who has spent a great deal of his life on his knees, though not in prayer. I do not say this last pridefully, but with the feeling that the posture, if not the thought behind it, may have had some final salutary effect. I am a naturalist and a fossil hunter, and I have crawled most of the way through life. I have crawled downward into holes without a bottom, and upward, wedged into crevices where the wind and the birds scream at you until the sound of a falling pebble is enough to make the sick heart lurch. In man, I know now, there is no such thing as wisdom. I have learned this with my face against the ground. It is a very difficult thing for a man to grasp today, because of his power; yet in his brain there is really only a sort of universal marsh, spotted at intervals by quaking green islands representing the elusive stability of modern science—islands frequently gone as soon as glimpsed.

It is our custom to deny this; we are men of precision, measurement and logic; we abhor the unexplainable and reject it. This, too, is a green island. We wish our lives to be one continuous growth in knowledge; indeed, we expect them to be. Yet well over a hundred years ago Kierkegaard observed that maturity consists in the discovery that "there comes a critical moment where everything is reversed, after which the point becomes to understand more and more that there is something which cannot be understood."

When I separated the serpent from the bird and released them in that wild upland, it was not for knowledge; not for anything I had learned in science. Instead, I contained, to put it simply, the serpent and the bird. I would always contain them. I was no longer one of the contending vapors; I had embraced them in my own substance and, in some insubstantial way, reconciled them. . . . I had transcended feather and scale and gone beyond them into another sphere of reality. I was trying to give birth to a different self whose only expression lies again in the deeply religious words of Pascal, "You would not seek me had you not found me."

I had not known what I sought, but I was aware at last that something had found me. I no longer believed that nature was either natural or unnatural, only that nature now appears natural to man. But the nature that appears natural to man is another version of the muskrat's world under the boat dock, or the elusive sparks over which the physicist made his trembling passage. They were appearances, specialized insights, but unreal because in the constantly onrushing future they were swept away.

What had become of the natural world of that gorilla-headed little ape from which we sprang—that dim African corner with its chewed fish bones and giant ice-age pigs? It was gone more utterly than my muskrat's tiny domain, yet it had given birth to an unimaginable thing—ourselves—something overreaching the observable laws of that far epoch. Man since the beginning seems to be awaiting an event the nature of which he does not know. "With reference to the near past," Thoreau once shrewdly commented, "we all occupy the region of common sense, but in the prospect of the future we are, by instinct, transcendentalists." This is the way of the man who makes nature "natural." He stands at the point where the miraculous comes into being, and after the event he calls it "natural." The imagination of man, in its highest manifestations, stands close to the doorway of the infinite, to the world beyond the nature that we know. Perhaps, after all, in this respect man constitutes the exertion of that act which Donne three centuries ago called God's Prerogative.

Man's quest for certainty is, in the last analysis, a quest for meaning. But the meaning lies buried within himself rather than in the void he has vainly searched for portents since antiquity. Perhaps the first act in its unfolding was taken by a raw beast with a fearsome head who dreamed some difficult and unimaginable thing denied his fellows. Perhaps the flashes of beauty and insight which trouble us so deeply are no less prophetic of what the race might achieve. All that prevents us is doubt—the power to make everything natural without the accompanying gift to see, beyond the natural, to that inexpressible realm in which the words "natural" and "supernatural" cease to have meaning.

Man, at last, is face to face with himself in natural guise. "What we make natural, we destroy," said Pascal. He knew, with superlative insight, man's complete necessity to transcend the worldly image that this word connotes. It is not the outward powers of man the toolmaker that threaten us. It is a growing danger which has already afflicted vast areas of the world—the danger that we have created an unbearable last idol for our worship. That idol, that uncreate and ruined visage which confronts us daily, is no less than man made natural. Beyond this replica of ourselves, this countenance already grown so distantly inhuman that it terrifies us, still beckons the lonely figure of man's dreams. It is a nature, not of this age, but of the becoming—the light once glimpsed by a creature just over the threshold from a beast, a despairing cry from the dark shadow of a cross on Golgotha long ago.

Man is not totally compounded of the nature we profess to understand. Man is always partly of the future, and the future he possesses a power to shape. "Natural" is a magician's word—and like all such entities, it should be used sparingly lest there arise from it, as now, some unglimpsed, unintended world, some monstrous caricature called into being by the indiscreet articulation of worn syllables. Perhaps, if we are wise, we will prefer to stand like those forgotten humble creatures who poured little gifts of flints into a grave. Perhaps there may come to us then, in some such moment, a ghostly sense that an invisible doorway has been opened—a doorway which, widening out, will take man beyond the nature that he knows.

Alan Watts
EMERSON'S "NAKED EYEBALL" REVISITED

Alan Watts is best known for his attempts to integrate Oriental religious thought and practice into American society. He has been closely identified with attempts to relate aspects of Zen Buddhism and Christian existentialism. He has also received recognition in recent years as a major interpreter of "alternative life-styles" in American life, particularly various forms of "consciousness-expansion." The following selection is no exception to Watt's interdisciplinary approach. It is an excerpt from The Book on the Taboo Against Knowing Who You Are, *on the problems of alienation and personal identity. In the following selection, taken from the chapter "The World Is Your Body," Watts seeks to integrate "Western pragmatism" and "Eastern intuition" to create a more comprehensive and penetrating humanism. The result seems to be a parallel development to the environmental philosophy of the biological sciences.*

The "nub" problem is the self-contradictory definition of man himself as a separate and independent being *in* the world, as distinct from a special action *of* the world. Part of our difficulty is that the latter view of man seems to make him no more than a puppet, but this is because, in trying to accept or understand the latter view, we are still in the grip of the former. To say that man is an action of the world is *not* to define him as a "thing" which is helplessly pushed

From *The Book,* by Alan Watts. Copyright © 1966 by Alan Watts. Reprinted by permission of Pantheon Books, a Division of Random House, Inc., and Jonathan Cape, Ltd.

around by all other "things." We have to get beyond Newton's vision of the world as a system of billiard balls in which every individual ball is passively knocked about by all the rest! Remember that Aristotle's and Newton's preoccupation with causal determinism was that they were trying to explain how one thing or event was influenced by others, forgetting that the division of the world into separate things and events was a fiction. To say that certain events are causally connected is only a clumsy way of saying that they are features of the same event, like the head and tail of the cat.

*   *   *

The illusion that organisms move entirely on their own is immensely persuasive until we settle down, as scientists do, to describe their behavior carefully. Then the scientist, be he biologist, sociologist, or physicist, finds very rapidly that he cannot say what the organism is doing unless, at the same time, he describes the behavior of its surroundings. Obviously, an organism cannot be described as walking just in terms of leg motion, for the direction and speed of this walking must be described in terms of the ground upon which it moves. Furthermore, this walking is seldom haphazard. It has something to do with food-sources in the area, with the hostile or friendly behavior of other organisms, and countless other factors which we do not immediately consider when attention is first drawn to a prowling ant. The more detailed the description of our ant's behavior becomes, the more it has to include such matters as density, humidity, and temperature of the surrounding atmosphere, the types and sources of its food, the social structure of its own species, and that of neighboring species with which it has some symbiotic or preying relationship.

When at last the whole vast list is compiled, and the scientist calls "Finish!" for lack of further time or interest, he may well have the impression that the ant's behavior is no more than its automatic and involuntary reaction to its environment. It is attracted by this, repelled by that, kept alive on one condition, and destroyed by another. But let us suppose that he turns his attention to some other organism in the ant's neighborhood—perhaps a housewife with a greasy kitchen—he will soon have to include that ant, and all its friends and relations, as something which determines *her* behavior! Wherever he turns his attention, he finds, instead of some positive, causal agent, a merely responsive hollow whose boundaries go this way and that according to outside pressures.

Yet, on second thought, this won't do. What does it *mean,* he asks himself, that a description of what the ant is doing must include what its environment is doing? It means that the thing or entity he is studying and describing has changed. It started out to be the individual ant, but it very quickly became the whole field of activities in which the ant is found. The same thing would happen if one started out to describe a particular organ of the body: it would be utterly unintelligible unless one took into account its relationships with other organs. It is thus that every scientific discipline for the study of living organisms—bacteriology, botony, zoology, biology, anthropology—must, from its own special standpoint, develop a science of ecology—literally, "the logic of the household"—or the study of organism/environment fields. Unfortunately, this science runs afoul of academic politics, being much too interdisciplinary for the jealous guardians of departmental boundaries. But the neglect of ecology is the one most serious weakness of modern technology, and it goes hand-in-hand with our reluctance to be participating members of the whole community of living species.

Man aspires to *govern* nature, but the more one studies ecology, the more absurd it seems to speak of any one feature of an organism, or of an organism/environment field, as governing or ruling others. Once upon a time the mouth, the hands, and the feet said to each other, "We do all this work gathering food and chewing it up, but that lazy fellow, the stomach, does nothing. It's high time he did some work too, so let's go on strike!" Whereupon they went many days without working, but soon found themselves feeling weaker and weaker until at last each of them realized that the stomach was *their* stomach, and that they would have to go back to work to remain alive. But even in physiological textbooks, we speak of the brain, or the nervous system, as "governing" the heart or the digestive tract, smuggling bad politics into science, as if the heart belonged to the brain rather than the brain to the heart or the stomach. Yet it is as true, or false, to say that the brain "feeds itself" through the stomach as that the stomach "evolves" a brain at its upper entrance to get more food.

*   *   *

To sum up: just as no thing or organism exists on its own, it does not act on its own. Furthermore, every organism is a process: thus the organism is not other than its actions. To put it clumsily: it is what it does. More precisely, the organism, including its behavior, is a process which is to be understood only in relation to the larger and longer process of its environment. For what we mean by "understanding" or "comprehension" is seeing how parts fit into a whole, and then realizing that they don't *compose* the whole, as one assembles a jigsaw puzzle, but that the whole is a pattern, a complex wiggliness, which *has* no separate parts. Parts are fictions of language, of the calculus of looking at the world through a net which *seems* to chop it up into bits. Parts exist only for purposes of figuring and describing, and as we figure the world out we become confused if we do not remember this all the time.

Once this is clear, we have shattered the myth of the Fully Automatic Universe where human consciousness and intelligence are a fluke in the midst of boundless stupidity. For if the behavior of an organism is intelligible only in relation to its environment, intelligent behavior implies an intelligent environment. Obviously, if "parts" do not really exist, it makes no sense to speak of an intelligent part of an unintelligent whole. It is easy enough to see that an intelligent human being implies an intelligent human society, for thinking is a social activity—a mutual interchange of messages and ideas based on such social institutions as languages, sciences, libraries, universities, and museums. But what about the non-human environment in which human society flourishes?

Ecologists often speak of the "evolution of environments" over and above the evolution of organisms. For man did not appear on earth until the earth itself, together with all its biological forms, had evolved to a certain degree of balance and complexity. At this point of evolution the earth "implied" man, just as the existence of man implies that sort of a planet at that stage of evolution. The balance of nature, the "harmony of contained conflicts," in which man thrives is a network of mutually interdependent organisms of the most astounding subtlety and complexity. Teilhard de Chardin has called it the "biosphere," the film of living organisms which covers the original "geosphere," the mineral planet. Lack of knowledge about the evolution of the organic from the "inorganic," coupled with misleading myths about life coming "into" this world from somewhere "outside," has made it difficult for us to see that the biosphere arises, or goes with, a certain degree of geological and astronomical evolution. But, as Douglas E. Harding has pointed out, we tend to think of this planet as a life-infested rock, which is as absurd as thinking of the human body as a cell-infested skeleton. Surely all forms of life, including man, must be understood as "symptoms" of the earth, the solar system, and the galaxy—in which case we cannot escape the conclusion that the galaxy is intelligent.

If I first see a tree in the winter, I might assume that it is not a fruit-tree. But when I return in the summer to find it covered with plums, I must exclaim, "Excuse me! You were a fruit-tree after all." Imagine, then, that a billion years ago some beings from another part of the galaxy made a tour through the solar system in their flying saucer and found no life. They would dismiss it as "Just a bunch of old rocks!" But if they returned today, they would have to apologize: "Well—you were peopling rocks after all!" You may, of course, argue that there is no analogy between the two situations. The fruit-tree was at one time a seed inside a plum, but the earth— much less the solar system or the galaxy—was never a seed inside a person. But, oddly enough, you would be wrong.

I have tried to explain that the relation between an organism and its environment is *mutual,* that neither one is the "cause" or deter- minant of the other since the arrangement between them is polar. If, then, it makes sense to explain the organism and its behavior in terms of the environment, it will also make sense to explain the environment in terms of the organism. (Thus far I have kept this up my sleeve so as not to confuse the first aspect of the picture.) For there is a very real, physical sense in which man, and every other organism, creates his own environment.

Our whole knowledge of the world is, in one sense, self-knowl- edge. For knowing is a translation of external events into bodily processes, and especially into states of the nervous system and the brain: we know the world *in terms* of the body, and in accordance with its structure. Surgical alterations of the nervous system, or, in all probability, sense-organs of a different structure than ours, give different types of perception—just as the microscope and telescope change the vision of the naked eye. Bees and other insects have, for example, polaroid eyes which enable them to tell the position of the sun by observing any patch of blue sky. In other words, because of the different structure of their eyes, the sky that they see is not the sky that we see. Bats and homing pigeons have sensory equipment analogous to radar, and in this respect see more "reality" than we do without our special instruments.

From the viewpoint of your eyes your own head seems to be an invisible blank, neither dark nor light, standing immediately behind the nearest thing you can see. But in fact the whole field of vision "out there in front" is a sensation in the lower back of your head, where the optical centers of the brain are located. What you see out there is, immediately, how the inside of your head "looks" or "feels." So, too, everything that you hear, touch, taste, and smell is some kind of vibration interacting with your brain, which translates that vibration into what you know as light, color, sound, hardness, roughness, saltiness, heaviness, or pungence. Apart from your brain, all these vibrations would be like the sound of one hand clapping, or of sticks playing on a skinless drum. Apart from your brain, or some brain, the world is devoid of light, heat, weight, solidity, motion, space, time, or any other imaginable feature. All these phenomena are interactions, or transactions, of vibrations with a certain arrangement of neurons. Thus vibrations of light and heat from the sun do not actually become light or heat until they interact with a living organism, just as no light-beams are visible in space unless reflected by particles of atmosphere or dust. In other words, it "takes two" to make anything happen. As we saw, a single ball in space has no motion, whereas two balls give the possibility of linear motion, three balls motion in a plane, and four balls motion in three dimensions.

The same is true for the activation of an electric current. No current will "flow" through a wire until the positive pole is connected with the negative, or, to put it very simply, no current will start unless it has a point of arrival, and a living organism is a "point of arrival" apart from which there can never be the "currents" or phenomena of light, heat, weight, hardness, and so forth. One might almost say that the magic of the brain is to evoke these marvels from the universe, as a harpist evokes melody from the silent strings.

*   *   *

. . . But what if it dawns on us that our perception of rocks, mountains, and stars is a situation of just the same kind? There is nothing in the least unreasonable about this. We have not had to drag in any such spooks as mind, soul, or spirit. We have simply been talking of an interaction between physical vibrations and the brain with its various organs of sense, saying only that creatures with brains are an *integral* feature of the pattern which also includes the solid earth and the stars, and that without this integral feature (or pole of the current) the whole cosmos would be as unmanifested as a rainbow without droplets in the sky, or without an observer. Our resistance to this reasoning is psychological. It makes us feel insecure because it unsettles a familiar image of the world in which rocks, above all, are symbols of hard, unshakable reality, and the Eternal Rock a metaphor for God himself. The mythology of the nineteenth century had reduced man to an utterly unimportant little germ in an unimaginably vast and enduring universe. It is just too much of a shock, too fast a switch, to recognize that this little germ with its fabulous brain is evoking the whole thing, including the nebulae millions of light-years away.

Does this force us to the highly implausible conclusion that before the first living organism came into being equipped with a brain there *was* no universe—that the organic and inorganic phenomena came into existence at the same temporal moment? Is it possible that all geological and astronomical history is a mere extrapolation— that it is talking about what *would* have happened *if* it had been observed? Perhaps. But I will venture a more cautious idea. The fact that every organism evokes its own environment must be corrected with the polar or opposite fact that the total environment evokes the organism. Furthermore, the total environment (or situation) is both spatial and temporal—both larger and longer than the organisms contained in its field. The organism evokes knowledge of a past before it began, and of a future beyond its death. At the other pole, the universe would not have started, or manifested itself, unless it was at some time going to include organisms—just as current will not begin to flow from the positive end of a wire until the negative terminal is secure. The principle is the same, whether it takes the universe billions of years to polarize itself in the organism, or whether it takes the current one second to traverse a wire 186,000 miles long.

I repeat that the difficulty of understanding the organism/environment polarity is psychological. The history and the geographical distribution of the myth are uncertain, but for several thousand years we have been obsessed with a false humility—on the one hand, putting ourselves down as mere "creatures" who came into this world by the whim of God or the fluke of blind forces, and on the other, conceiving ourselves as separate personal egos fighting to control the physical world. We have lacked the real humility of recognizing that we are members of the biosphere, the "harmony of contained conflicts" in which we cannot exist at all without the cooperation of plants, insects, fish, cattle, and bacteria. In the same measure, we have lacked the proper self-respect of recognizing that I, the individual organism, am a structure of such fabulous ingenuity that it calls the whole universe into being. In the act of putting everything at a distance so as to describe and control it, we have orphaned ourselves both from the surrounding world and from our own bodies—leaving "I" as a dis*content*ed and alienated spook, anxious, guilty, unrelated, and alone.

We have attained a view of the world and a type of sanity which is dried-out like a rusty beer-can on the beach. It is a world of *objects,* of nothing-buts as ordinary as a formica table with chromium fittings. We find it immensely reassuring—except that it won't stay put, and must therefore be defended even at the cost of scouring the whole planet back to a nice clean rock. For life is, after all, a rather messy and gooey accident in our basically geological universe. "If a man's son ask for bread, will he give him a stone?" The answer is probably "Yes."

Yet this is no quarrel with scientific thinking, which, as of this date, has gone far, far beyond Newtonian billiards and the myth of the Fully Automatic, mechanical universe of mere objects. That was where science really got its start, but in accordance with William Blake's principle that "The fool who *persists* in his folly will become wise," the persistent scientist is the first to realize the obsolescence of old models of the world.

Theodore Roszak
TRUSTING THE WAY OF THE EARTH

In place of dominant scientific, political, military, or economic interests today, Theodore Roszak believes that American civilization must be re-newed by a very different "visionary imagination." He explains that it was once the shaman, the medicine man of primitive societies, who called for total human fulfillment by means of a symbiotic relationship between man and "not-man." Such a relationship would secure for man his true dignity, gracefulness, and intelligence. He would free himself from a surface con-sciousness and gain access to hidden powers in the universe. Historian at California State College at Hayward, Roszak is a leading interpreter of "counter-culture" alternatives to existing trends in American life. He is particularly critical of the limits of "objective consciousness," or scientific and objective ways of experiencing the world. He believes they force upon men a false separation between objective and subjective knowledge. They cause human alienation and mechanize knowledge. In the following selec-tion, from the last chapter in his book, Roszak includes ecological sensitivity in his refutation of existing "counterfeit Nature" fabricated out of scientific ways. He praises the subjective, symbiotic, and organic aspects of man and the world. Roszak has been criticized for supporting a "tender-minded" romantic primitivism, escapism, and anti-intellectualism. But he argues that his "counter culture" is far more attractive than a coercive technological society that destroys men's souls, despoils their landscape, and manipulates its citizens.

> *"What," it will be Question'd, "When the Sun rises, do you not see a round disk of fire somewhat like a Guinea?" O no, no, I see an Innumer-able company of the Heavenly host crying, "Holy, Holy is the Lord God Almighty."*
>
> *—William Blake*

What are we to say of the man who fixes his eye on the sun and

From *The Making of a Counter Culture* by Theodore Roszak; copyright © 1968, 1969 by Theodore Roszak. Reprinted by permission of Doubleday & Company, Inc., and Faber and Faber Ltd. Notes to the original have been omitted.

does not see the sun, but sees instead a chorus of flaming seraphim announcing the glory of God? Surely we shall have to set him down as mad . . . unless he can coin his queer vision into the legal tender of elegant verse. Then, perhaps, we shall see fit to assign him a special status, a pigeonhole: call him "poet" and allow him to validate his claim to intellectual respectability by way of metaphorical license. Then we can say, "He did not *really* see what he says he saw. No, not at all. He only put it that way to lend color to his speech . . . as poets are in the professional habit of doing. It is a lyrical turn of phrase, you see: just that and nothing more." And doubtless all the best, all the most objective scholarship on the subject would support us in our perfectly sensible interpretation. It would tell us, for example, that the poet Blake, under the influence of Swedenborgian mysticism, developed a style based on esoteric visionary correspondences and was, besides, a notorious, if gifted, eccentric. Etc. Etc. Footnote.

In such fashion, we confidently discount and denature the visionary experience, and the technocratic order of life rolls on undeterred, obedient to the scientific reality principle. From such militant rationality the technocracy must permit no appeal.

Yet, if there is to be an alternative to the technocracy, there *must* be an appeal from this reductive rationality which objective consciousness dictates. This, so I have argued, is the primary project of our counter culture: to proclaim a new heaven and a new earth so vast, so marvelous that the inordinate claims of technical expertise must of necessity withdraw in the presence of such splendor to a subordinate and marginal status in the lives of men. To create and broadcast such a consciousness of life entails nothing less than the willingness to open ourselves to the visionary imagination on its own demanding terms. We must be prepared to entertain the astonishing claim men like Blake lay before us: that here are eyes which see the world not as commonplace sight or scientific scrutiny sees it, but see it transformed, made lustrous beyond measure, and in seeing the world so, see it as it really is. Instead of rushing to downgrade the rhapsodic reports of our enchanted seers, to interpret them at the lowest and most conventional level, we must be prepared to consider the scandalous possibility that wherever the visionary imagination grows bright, magic, that old antagonist of science, renews itself, transmuting our workaday reality into something bigger, perhaps more frightening, certainly more adventurous than the lesser rationality of objective consciousness can ever countenance.

But to speak of magic is to summon up at once images of vaudeville prestidigitators and tongue-in-cheek nature-fakers: tricksters who belong to the tawdry world of the stage. We have learned in this enlightened age to tolerate magicians only as an adjunct of the entertainment industry, where it is strictly understood by performer and audience alike that a trick is no more than a trick, a practised effort to baffle us. When the impossible appears to happen on stage, we know better than to believe that it has really happened. What we applaud is the dexterity with which the illusion has been created. If the magician were to claim that his deed was more than an illusion, we would consider him a lunatic or a charlatan, for he would be asking us to violate our basic conception of reality; and this we would not tolerate. While there are many, surprisingly many, who remain willing to take spiritualists, faith healers, fortunetellers, and such seriously, the scientific skeptic is forced to discount all these phenomena as atavistic and to insist stubbornly on the primacy of a coherent world view. The skeptical mind argues doggedly that we live in the midst of a nature that has been explained and exploited by science. The vaccines we inject into our bodies, the electricity that goes to work for us at the flick of a switch, the airplanes and automobiles that transport us: these and the ten thousand more technological devices we live among and rely upon derive from the scientist's, not the charlatan's, conception of nature. How shall we, with intellectual conscience, enjoy so much of what science has with an abundance of empirical demonstration brought us, and then deny the essential truth of its world view?

*   *   *

*The tongues of the Lightning Snakes flicker and twist, one to the
    other . . .
They flash across the foliage of the cabbage palms . . .
Lightning flashes through the clouds; with the flickering tongues
    of the Snake . . .*

> All over the sky, their tongues flicker: at the place of the Two
>   Sisters, the place of the Wauwalak
> Lightning flashes through the clouds, flash the Lightning Snake . . .
> Its blinding flash lights up the cabbage palm foliage . . .
> Gleams on the cabbage palms and on the shining leaves . . .*

Now, to see the world in this way is precisely what our culture is prone to call "superstition." We are forced to interpret the fact that the human race survived by such an understanding of nature for tens of thousands of years as so much dumb luck. To believe that this magical vision is anything but a bad mistake or, at best, a primitive adumbration of science, is to commit heresy. And yet from such a vision of the environment there flows a symbiotic relationship between man and not-man in which there is a dignity, a gracefulness, an intelligence that powerfully challenges our own strenuous project of conquering and counterfeiting nature. From that "superstitious" perception, there derives a sense of the world as our house, in which we reside with the ease, if not always the comfort, of creatures who trust the earth that raised them up and nurtures them.

The trouble is, we *don't* trust to the way of the world. We have learned—in part from the accelerating urbanization of the race, in part from the objective mode of consciousness so insistently promulgated by Western science, in part, too, perhaps from the general Christian disparagement of nature—to think of the earth as a pit of snares and sorrows. Nature is that which must be taken unsentimentally in hand and made livable by feverish effort, ideally by replacing more and more of it with man-made substitutes. So then, perhaps someday we shall inhabit a totally plastic world, clinically immaculate and wholly predictable. To live in such a completely programmed environment becomes more and more our conception of rational order, of security. Concomitantly, our biologists begin to think even of the genetic process as a kind of "programming" (though, to be sure, a faulty one that can be improved upon in a multitude of ways). The object almost seems to bear out the ideas

---

* R. M. and C. H. Berndt, *World of the First Australians* (Chicago: University of Chicago Press, 1965), p. 315. Reprinted by permission.

of Otto Rank's return-to-the-womb psychology, with our goal being a world-wide, lifelong plastic womb. . . .

As a culture, we have all but completely lost the eyes to see the world in any other way. In contrast to the hard-edged, distinct focus of the scientist's impersonal eye, which studies this or that piece of the environment in order to pry its secrets from it, the sensuous, global awareness of the shaman seems like that sort of peripheral vision which is intolerably imprecise. Our habit is to destroy this receptive peripheral vision in favor of particularistic scrutiny. We are convinced that we learn more in this way about the world. And, after a fashion, we do learn things by treating the world objectively. We learn what one learns by scrutinizing the trees and ignoring the forest, by scrutinizing the cells and ignoring the organism, by scrutinizing the detailed minutiae of experience and ignoring the whole that gives the constituent parts their greater meaning. In this way we become ever more learnedly stupid. Our experience dissolves into a congeries of isolated puzzles, losing its overall grandeur. We accumulate knowledge like the miser who interprets wealth as maniacal acquisition plus tenacious possession; but we bankrupt our capacity to be wonderstruck . . . perhaps even to survive.

Consider for a moment the admonition of the quaint old Wintu woman, who warns that the "spirit of the earth" hates us for what we have done to our environment. Of course we *know* there is no "spirit of the earth." But even now as I write and as you read, there reside in the bowels of the earth, in concrete silos throughout our advanced societies, genocidally destructive weapons capable of annihilating our safe and secure civilization. No doubt in her deeply poetic imagination the old woman would see in these dread instruments the vengeful furies of the earth poised to repay the white man for his overweening pride. A purely fanciful interpretation of our situation, we might say. But maybe there is more truth in the old woman's poetry than in our operations analysis. Maybe she realizes that the spirit of the earth moves in more mysterious ways than we dare let ourselves believe, borrowing from man himself its instruments of retribution.

I have argued that the scientific consciousness depreciates our capacity for wonder by progressively estranging us from the magic of the environment. Is the charge unfair to science? Do not scientists, like the visionary poets, also teach us of the "beauties" and "wonders" of nature?

To be sure, they appropriate the words. But the experience behind the words is not the same as that of the shamanistic vision. The mode of objective consciousness does not expand man's original sense of wonder. Rather, it displaces one notion of beauty by another, and, in so doing, cuts us off from the magical sense of reality by purporting to supersede it. The beauty which objective consciousness discerns in nature is that of generalized orderliness, of formal relationships worked out by In-Here as it observes things and events. This is the beauty of the efficiently solved puzzle, of the neat classification. It is the beauty a chess player discovers in a well-played game or a mathematician in an elegant proof. Such nomothetic beauties are conveniently summed up and indeed certified by a formula or a diagram or a statistical generalization. They are the beauties of experience planed down to manageable and repeatable terms, packaged up, mastered, brought under control. In accordance with the ideal of scientific progress, such beauties can be salted away in textbooks and passed onto posterity in summary form as established conclusions.

In contrast, the beauty of the magical vision is the beauty of the deeply sensed, sacramental presence. The perception is not one of order, but of power. Such experience yields no sense of accomplished and rounded-off knowledge, but, on the contrary, it may begin and end in an overwhelming sense of mystery. We are awed, not informed. The closest most of us are apt to come nowadays to recapturing this mode of experience would be in sharing the perception of the poet or painter in the presence of a landscape, of the lover in the presence of the beloved. In the sweep of such experience, we have no interest in finding out about, summing up, or solving. On the contrary, we settle for celebrating the sheer, amazing fact that this wondrous thing is self-sufficiently there before us. We lose ourselves in the splendor or the terror of the moment and ask no more. We leave what we experience—this mountain, this sky, this place filled with forbidding shadows, this remarkable person—to be what it is, for its being alone is enough.

The scientist studies, sums up, and has done with his puzzle; the painter paints the same landscape, the same vase of flowers, the same person over and over again, content to reexperience the inexhaustible power of this presence interminably. The scientist reduces the perception of colored light to a meteorological generalization; the intoxicated poet announces, "My heart leaps up when I behold a rainbow in the sky," and then goes on to find a hundred ways to say the same thing over again without depleting the next poet's capacity to proclaim the same vision still again. What conceivable similarity is there between two such different modes of experience? None whatsoever. One clichéd argument suggests that the work of the scientist *begins* with the poet's sense of wonder (a dubious hypothesis at best) but then goes *beyond* it armed with spectroscope and light meter. The argument misses the key point: the poet's experience is defined precisely by the fact that the poet does *not* go beyond it. He begins and ends with it. Why? Because it is sufficient. Or rather, it is inexhaustible. What he has seen (and what the scientist has *not* seen) is not improved upon by being pressed into the form of knowledge. Or are we to believe it was by failure of intelligence that Wordsworth never graduated into the status of weatherman?

* * *

It would be one of the bleakest errors we could commit to believe that occasional private excursions into some surviving remnant of the magical vision of life—something in the nature of a psychic holiday from the dominant mode of consciousness—can be sufficient to achieve a kind of suave cultural synthesis combining the best of both worlds. Such dilettantism would be a typically sleazy technocratic solution to the problem posed by our unfulfilled psychic needs; but it would be a deception from start to finish. We have either known the magical powers of the personality or we have not. And if we have felt them move within us, then we shall have no choice in the matter but to liberate them and live by the reality they illuminate. One does not free such forces on a part-time basis any more than one falls madly in love or repents of sin on a part-time basis. To suggest that there may be some halfway house between the magical and the objective consciousness in which our culture can reside is quite simply to confess that one does not know what it is to see with the eyes of fire. In which case, we shall never achieve the personal, transactive relationship with the reality that envelops us which is the essence of the magical world view. Accordingly, whatever our degree of intellectual sophistication, we shall as a culture continue to deal with our natural environment as lovingly, as reverently as a butcher deals with the carcass of a dead beast.

Yet, if we have lost touch with the shamanistic world view by which men have lived since the paleolithic beginnings of human culture, there is one sense in which magic has not lost its power over us with the progress of civilization. It is not only the dumfounded populations of so-called underdeveloped societies that perceive and yield to the white man's science and technology as a form of superior magic. The same is true of the white man's own society—though we, as enlightened folk, have learned to take the magic for granted and to verbalize various non-supernatural explanations for its activity. True enough: science possesses theory, methodology, epistemology to support its discoveries and inventions. But, alas! most of us have no better understanding of these things than the bewildered savages of the jungle. Even if we have acquired the skill to manipulate vacuum tubes and electrical circuits and balky carburetors, few of us could articulate one commendable sentence about the basic principles of electricity or internal combustion, let alone jet propulsion, nuclear energy, deoxyribonucleic acid, or even statistical sampling, which is supposedly the key to understanding our own collective opinions these days.

It is remarkable how nonchalantly we carry off our gross ignorance of the technical expertise our very lives depend upon. We live off the surface of our culture and pretend we know enough. If we are cured of disease, we explain the matter by saying a pill or a serum did it—as if that were to say anything at all. If the economy behaves erratically, we mouth what we hear about inflationary pressures . . . the balance of payments . . . the gold shortage . . . down-turns and up-swings. Beyond manipulating such superficial notions, we work by faith. We believe that somewhere behind the pills and the economic graphs there are experts who understand whatever else there is to understand. We know they are experts, because, after all, they talk like experts and besides possess degrees, licenses, titles, and certificates. Are we any better off than the savage who believes his fever has been cured because an evil spirit has been driven out of his system?

For most of us the jargon and mathematical elaborations of the experts are so much mumbo jumbo. But, we feel certain, it is all mumbo jumbo that *works*—or at least seems to work, after some fashion that the same experts tell us should be satisfactory. If those who know best tell us that progress consists in computerizing the making of political and military decisions, who are we to say this is not the best way to run our politics? If enough experts told us that strontium 90 and smog were good for us, doubtless most of us would take their word for it. We push a button and something called the engine starts; we press a pedal and the vehicle moves; we press the pedal more and it moves faster. If we believe there is someplace to get and if we believe it is important to get there very, very fast— despite the dangers, despite the discomforts, despite the expense, despite the smog—then the automobile is an impressive piece of magic. That is the sort of magic science can bring about and which shamanistic incantations never will. Push another button and the missile blasts off; aim it correctly and it will blow up a whole city . . . maybe, if the hardware is sophisticated enough, the whole planet. If blowing up the planet is deemed worth doing (under certain well-considered conditions, to be sure), then science is what we want. Incantations will never do the job.

But if the role of the technical expert in our society is analogous to that of the old tribal shaman—in the sense that both are deferred to by the populace at large as figures who conjure with mysterious forces in mysterious ways—what significant difference is there between cultures based on scientific and visionary experience? The difference is real and it is critical. It requires that we make a distinction between good and bad magic—a line that can be crossed in any culture, primitive or civilized, and which has been crossed in ours with the advent of the technocracy.

* * *

In harking back to the shamanistic world view, a cultural stage buried in the primitive past of our society, I may seem to have strayed a long way from the problems of our contemporary dissenting youth. But that is hardly the case. The young radicalism of our day gropes toward a critique that embraces ambitious historical and comparative cultural perspectives. The New Left that rebels against technocratic manipulation in the name of participative democracy draws, often without realizing it, upon an anarchist tradition which has always championed the virtues of the primitive band, the tribe, the village. The spirit of Prince Kropotkin, who learned the anti-statist values of mutual aid from villagers and nomads little removed from the neolithic or even paleolithic level, breathes through all the young say about community. Our beatniks and hippies press the critique even further. Their instinctive fascination with magic and ritual, tribal lore, and psychedelic experience attempts to resuscitate the defunct shamanism of the distant past. In doing so, they wisely recognize that participative democracy cannot settle for being a matter of political-economic decentralism—only that and nothing more. As long as the spell of the objective consciousness grips our society, the regime of experts can never be far off; the community is bound to remain beholden to the high priests of the citadel who control access to reality. It is, at last, reality itself that must be participated in, must be seen, touched, breathed with the conviction that *here* is the ultimate ground of our existence, available to all, capable of ennobling by its majesty the life of every man who opens himself. It is participation of this order—experiential and not merely political—that alone can guarantee the dignity and autonomy of the individual citizen. The strange youngsters who don cowbells and primitive talismans and who take to the public parks or wilderness to improvise outlandish communal ceremonies are in reality seeking to ground democracy safely beyond the culture of expertise. They give us back the image of the paleolithic band, where the community during its rituals stood in the presence of the sacred in a rude equality that predated class, state, status. It is a strange brand of radicalism we have here that turns to prehistoric precedent for its inspiration.

To be sure, there is no revolutionizing the present by mere reversion to what is for our society a remote past. Prehistoric or contemporary primitive cultures may serve as models to guide us; but they can scarcely be duplicated by us. As Martin Buber warns us in his discussion of the magical world view of primitive man, "he who attempts a return ends in madness or mere literature." It is, as he says, a *"new* pansacramentalism" we need, one which works within and expands the interstices of the technocracy, responding wherever possible to the thwarted longings of men. There will have to be experiments—in education, in communitarianism—which will seek not coexistence with the technocracy and less still the treacherous satisfactions of quick publicity; but which aim instead at subverting and seducing by the force of innocence, generosity, and manifest happiness in a world where those qualities are cynically abandoned in favor of bad substitutes. To the end that there shall be more and more of our fellows who cease to live by the declared necessities of the technocracy; who refuse to settle for a mere after-hours outlet for the magical potentialities of their personalities; who become as if deaf and blind to the blandishments of career, affluence, the mania of consumption, power politics, technological progress; who can at last find only a sad smile for the low comedy of these values and pass them by.

But further, to the end that men may come to view much that goes by the name of social justice with a critical eye, recognizing the way in which even the most principled politics—the struggle against racial oppression, the struggle against world-wide poverty and backwardness—can easily become the lever of the technocracy in its great project of integrating ever more of the world into a well-oiled, totally rationalized managerialism. In a sense, the true political radicalism of our day begins with a vivid realization of how much in the way of high principle, free expression, justice, reason, and humane intention the technocratic order can adapt to the purpose of entrenching itself ever more deeply in the uncoerced allegiance of men. This is the sort of insight our angriest dissenters tend to miss when, in the course of heroic confrontation, they open themselves to the most obvious kinds of police and military violence. They quickly draw the conclusion that the status quo is supported by nothing more than bayonets, overlooking the fact that these bayonets enjoy the support of a vast consensus which has been won for the status quo by means far more subtle and enduring than armed force.

For this reason, the process of weening men away from the technocracy can never be carried through by way of a grim, hard-bitten, and self-congratulatory militancy, which at best belongs to tasks of ad hoc resistance. Beyond the tactics of resistance, but shaping them at all times, there must be a stance of life which seeks not simply to muster power against the misdeeds of society, but to transform the very sense men have of reality. This may mean that, like George Fox, one must often be prepared not to act, but to "stand still in the light," confident that only such a stillness possesses the eloquence to draw men away from lives we must believe they inwardly loathe, but which misplaced pride will goad them to defend under aggressive pressure to the very death—their death and ours.

Eric A. Walker

TECHNOLOGY CAN BECOME MORE HUMAN

Eric A. Walker, president of Pennsylvania State University, has had a strong influence on the directions of scientific education in the United States. He is the chief editor of the Goals Report of the American Society for Engineering Education. While long an advocate of scientific and technological development as a primary means to improve human life, Walker has recently acknowledged "deterioration in the quality of our lives." But unlike most other authors in this volume, he is confident in America's inherent ability as a nation to overcome squalor and despair by engineering means. Today, however, Walker admits that engineering and scientific leadership in the past paid little attention to its aesthetic, moral, and environmental responsibilities. This neglect of the human factor has been destructive and accounts for much of today's ecological crisis. He calls for a massive marshaling of the nation's material and human resources, led jointly by the humanities, social sciences, and natural sciences, each to inform and assist the other. Walker believes that a revitalized educational system, more closely related to the needs of society, is the key to a future America more responsive to its people and its environment.

From "Engineers and the Nation's Future," by Eric A. Walker, published in *Approaching the Benign Environment,* edited by Taylor Littleton. Copyright © 1970 by The University of Alabama Press. Reprinted by permission.

The rapid and extensive growth of research, in industry, in educational institutions, and within the government itself, has tended to pull the scientist and the engineer closer together, and blur the differences in their separate approaches to the requirements of modern life. The engineer, in his design of new and sophisticated devices, has become increasingly dependent upon the newly found knowledge of the scientist, and does more research himself. And the scientist himself has frequently found it necessary to work hand-in-hand with the engineer and at times be a designer and engineer.

So today, in the public's mind, there seems to be no clear-cut distinction between the activities of the scientist and those of the engineer. Yet the distinction does exist.

In his book *Two Cultures and the Scientific Revolution,* C. P. Snow pointed out that a great gulf exists between this science-engineering complex on the one hand, and the rest of the population—particularly, he says, the literary intellectuals—on the other. Snow blames this gulf on the lack of communication between the two groups and he says that it well could be fatal to the Western world.

Yet, the gulf is deeper, broader, and somewhat different from that, and the real problem is not in communications as such. The real gulf today, I believe, lies between what science and engineering are capable of doing for mankind on the one hand, and what the average citizen is getting as a result of all this knowledge, on the other.

There was a time, up until a few decades ago, when we were a nation of practical doers who could put together machines and do almost anything. We developed the telegraph, the telephone, the sewing machine, and the cotton gin. Men like Edison came along and provided a score of practical inventions for this nation and for the world. Almost anything engineers did to harness nature then was considered good. The engineer was the builder of the highways, bridges, skyscrapers, and the designer of the industrial machinery and the processes that made our factories grow. And at first, if the highways got a little crowded, or the fumes from the factory darkened the skies, or if the tightly packed skyscrapers crowded too many people together in one spot—well, that was considered the necessary price of progress.

During those years, we did almost nothing about pure science. We believed and demonstrated that necessity was the mother of invention. We knew how to make good Kentucky rifles before we knew anything about the science of metallurgy. And steam engines worked successfully and provided power before we understood the laws of thermodynamics. Yet as a result of our inventiveness, our ingenuity, and our practical know-how, we built up an economic base that enabled us to support pure science, art, music, and literature.

*   *   *

Until a few years ago, we seem to have made the assumption that so long as we provided enough funds for basic research, the new knowledge we discovered would almost automatically be translated into products and systems and services that would enrich our lives and create general prosperity. And indeed there is evidence that in some instances this sort of transfer occurs quickly. Yet in the past few years, we have had reason to question how widespread this process of transfer really is. Many of our Congressmen and other public officials have begun to wonder whether the fantastic sums of public money that are being poured into basic research are really paying off in terms of practical progress. In many instances it would appear that knowledge is accumulating at such a rate and to an extent that it cannot possibly be used effectively without a more conscious effort to put it to practical use for the good of humanity.

We are filling our libraries with an almost unbelievable amount of new knowledge, and we ask ourselves what practical use can be made of all these facts, theories, and discoveries. Is the growth of our funding of basic research disproportionate to the growth of our gross national product? Can we continue to give an ever increasing share of our GNP to science? If so, who gets less—welfare, health care, highways, or old age pensions? Should we not examine more critically the whole process of innovation and invention, the means by which basic knowledge is actually applied to practical use? Have we failed to give proper attention to this vital step in the process?

The trouble is that, while we have expended a good deal of money on pure science, we have not done enough on the engineering end of the problem for our people. Now once again Americans are beginning to ask the question: "Of what value is pure science to me?"

It is my belief that the average citizen in America today looks around him and is not satisfied with what science and engineering have given him. Let me rephrase that and say that the public is not satisfied mainly because there are many things that could be done for the citizen which are not being done by the scientists and engineers who are using public money. Today we are witnessing a great deterioration in the quality of our lives, caused principally, I believe, because of this situation. For example, we have gone to the moon, and here on earth we have very fine automobiles, very fine airplanes, and in many places very fine highways. But the average citizen still can't get from one place to another rapidly and safely.

*   *   *

Chauncey Starr, dean of engineering at UCLA, says in a recent publication that many of our environmental pollution problems have presently known engineering solutions—but the problems of economic readjustment, political jurisdiction, and social behavior loom as very large obstacles. If we continue on the current path, he says, it will take many decades to put into practice the technical solutions we know today. As a specific illustration, the pollution of our water resources is completely avoidable by engineering systems now available; but, in fact, interest in making the economic and political adjustments to apply these techniques is very limited. In most cases town fathers just don't want to take the risk of trying new and expensive systems. And I wouldn't either if my reelection hung on the failure of an engineering experiment. It has been facetiously suggested that, as a means of motivating people, every community and industry should be required to place its drinking water intake downstream from its sewage discharges.

*   *   *

It is only fair to say, of course, that there is evidence that we have begun to recognize the housing challenge, at least in some quarters. For example, the National Academy of Engineering is talking about a proposal put together some time ago by the President's Office of Science and Technology for an Interagency Family Housing Demonstration Program. This would leap over local codes, union rules, and antiquated assembly processes to build good houses inexpensively.

Recognizing that real innovation in the construction of housing units is technologically possible and highly desirable, but that development of new techniques has been hampered by the fragmentation of the industry, code constraints, rigid labor practices, and government apathy, the OST proposed an extensive project, to be undertaken jointly by the Department of Housing and Urban Development, and the Department of Defense. The idea was to start from scratch, so to speak, by securing a large area outside the jurisdiction of local building codes or other restrictions, and attack the problem of family housing in much the same way that a large and complicated weapons system is attacked, through a program advancing from small-scale field experimentation and evaluation to prototypes, to large-scale procurement for demonstration purposes. Every possible use is to be made of new techniques and innovations. It is expected that by using government construction funds to demonstrate the practicality of advanced building systems, the stage could be set for substantial improvement in low-income housing nationally. Projects of this sort, I think, are badly needed. For it is clear to me that if improvements don't come in all these situations we will reach a time when people will demand that something be done. And after all, it is the average citizen who is paying the bill in this country.

It is my feeling that these problems, taken together, are as severe as they were when we were menaced by enemies during the first and second World Wars. We all know that during those wars, the nation's and the people's entire energies were devoted to overcoming our problems. We turned our entire attention to the practical aspects of science and engineering to get something done—something that was needed immediately in the face of clear and present danger. We mobilized all our resources. I believe that we are again in the middle of a crisis—a clear and present danger—although many of us don't recognize it as that yet.

Perhaps what must be done again is that the whole country should be mobilized to solve these national problems—government, industry, the public, pure scientists, applied scientists, engineers, and businessmen. Perhaps we could have a "call to arms" in which the Federal and State governments recognize that the solution to these public problems is important and crucial. Perhaps with the end of the Vietnam War we can turn our attention—our whole attention and not partial attention—to the solution of such problems. Perhaps there could be a four-year period in which the whole country "wars" on the deterioration of the quality of our life. This would take mobilization of all our efforts, from the highest level on down, including the Department of Defense, the Space Agency, the National Science Foundation, the National Academy of Engineering, and all our universities.

*　　　*　　　*

There is no doubt in my mind that if we turn our attention now to practical problems and are willing to wage a war against the deterioration in the quality of our lives—even for as short a period as four years—the advance we could achieve for humanity would be phenomenal. And it would set us apart as a nation for all to follow— a nation which is willing to seriously approach the problems of its citizens and find the solutions for them. To do so would not be a difficult task when one considers that approximately 90 percent of all the scientists and engineers who ever lived are alive today.

Certainly a large share of the blame for our slowness to act on these national problems must be laid to technology itself—to the engineer who has traditionally confined his interests all too narrowly to his technical specialty, and who is only beginning to take his proper place in the mainstream of social activity. It seems to me that, more than any other group, the engineering profession must accept this new responsibility wholeheartedly.

The world of the engineer can no longer be limited to the concerns of business and industry, or even to the traditional requirements in the field of public works. Along with the rest of society, the engineer is faced with the task of finding solutions to these pressing problems of modern life—problems which lie in the public sector of our economy. Basically, many of these problems are engineering problems. To a large extent, their ultimate solutions will depend upon the willingness and ability of the engineering community to provide society with the kind of help and advice that is needed to solve them.

Engineering must, and is, more and more recognizing the importance of the social sciences, the humanities, and the communication skills in the undergraduate programs. It is evident that the engineer of the future will be called upon more and more to play an increasingly active role in the solution of complex social problems. He will have to cope not only with the physical forces, as in the past, but with biological, social, and political forces. As a consequence, engineering education in the years ahead—as well as science education—will have to impart a thorough knowledge of the many nontechnical aspects of modern life. And the humanities, too, must begin to learn more and more about the professions of science and engineering, and must learn to work more closely with their colleagues in these fields.

Just as our government has financed big science, it must now finance big engineering. But the engineers must prove that they are able to take government money to solve the people's problems and give them what they need to be productive, efficient, and contented citizens. To make a start in solving these national problems would bring us a great deal more prestige as a nation—prestige that we have been losing for the past several years.

Lincoln reminded us that a house divided against itself cannot long stand. So too, a nation that devotes a great deal of its energies and resources to pure science without an equal return in practical benefits for the citizens who are paying the bill, cannot hope to have the support of its citizens. We must not forget, as we reach out into the universe, that this government was founded to "insure domestic tranquillity" and to "promote the general welfare"—right here on earth.

Herman Kahn and Anthony J. Wiener
MAN'S FAUSTIAN POWERS

Herman Kahn and Anthony J. Wiener are leading members of the Hudson Institute in New York State. As such, their writings represent statements of advanced political, social, and economic thought in American government and business. The Hudson Institute itself, like the Rand Corporation, the Brookings Institution, and other "think tanks," is one of the most influential professional communities of scholars and scientists, planners and prophets, ever to exist in American history. Government policies on national and international levels are often deeply affected by the opinions of Kahn, Wiener, and their colleagues. To their critics, they epitomize the Establishment in American society. Their writings about the future of technological civilization are thus of considerable interest in environmental terms. The authors say they are aware of the incredibly extensive and complicated problems created by industrial society, but they are ultimately hopeful, even confident, of a more successful future humanity by the same means.

To speak, to use tools, to pass learning on to children; to put fire, domestic animals, wind, falling water, and other energy sources to human use; to gather food, fuel, clothing, and seeds for winter; to save, invest, plan, build, and innovate in order to decrease dangers and insecurities and to increase the power to change natural things to suit one's purposes; in sum, to subdue Nature, and render her subject to human will—such have been the results, if not always the conscious goals, of eons of striving. Success would seem to be at hand; as we approach the beginning of the twenty-first century, our capacities for and commitment to economic development and technological control over our external and internal environment, as well as the concomitant systematic innovation, application, and diffusion of these capacities, seem to be increasing, and without foreseeable limit.

Clearly it is worthwhile to overcome both the deprivations caused by economic scarcity and the dangers and frustrations caused by impotence before the forces of nature. To increase economic de-

From "Faustian Powers and Human Choices: Some Twenty-First Century Technological and Economic Issues," by Herman Kahn and Anthony J. Wiener, in *Environment and Change: The Next Fifty Years* (Bloomington, Ind.: Indiana University Press, 1968). Copyrighted by the Hudson Institute and reprinted by their permission. Notes to the original have been omitted.

185

velopment is to increase the availability of at least some of the things that people need and want. To develop technologically is to increase the capacity to achieve at least some human purposes that are widespread and legitimate. These Promethean accomplishments, though they are mixed blessings, are the results of persistent and concerted effort and intelligence, and on the whole they are occasions for satisfaction.

In this case, as in others, it has been desirable to solve old problems in spite of the new problems created by the solutions themselves. The purpose of this paper, however, is to focus attention on some of the new problems created by technological and economic progress. Through such progress such issues arise as the accumulation, augmentation, and proliferation of weapons of mass destruction; the loss of privacy and solitude; the increase of governmental and/or private power over individuals; the loss of human scale and perspective and the dehumanization of social life or even of the psychobiological self; the growth of dangerously vulnerable, deceptive or degradable centralization of administrative or technological systems; the creation of other new capabilities so inherently dangerous as seriously to risk disastrous abuse; and the acceleration of changes that are too rapid or cataclysmic to permit successful adjustment. Perhaps most crucial, choices are posed that are too large, complex, important, uncertain, or comprehensive to be safely left to fallible humans, whether they are acting privately or publicly, individually or in organizations—choices, however, that become inescapable once these new capabilities have been gained.

The capacities of our culture and institutions to adapt to so much change in so comparatively short a time may be a major question; the stresses in domestic societies and in the international system may not be managed sufficiently by meliorist policies. Since the underdeveloped countries are even further removed in industrial and social life from these new technologies than we are, the cultural shock of their partial adaptation to the new technologies may even be greater for them. The possibility must be faced that man's unremitting, Faustian striving may ultimately remake his natural conditions—environmental, social, and psychobiological—so far as to begin to dehumanize man or to degrade his political or ecological situation in very costly or even irrevocable ways.

We would not wish to arrest this process, perhaps not even to slow it down; on balance, it is too valuable. There is a widespread though simplistic, Luddite response to problems of technology, in which the artifacts themselves become the targets of hostility; surely it would be generally more useful to criticize specific human choices, in which technology plays but a passive, instrumental role. Nor do we find plausible the prevalent rhetoric to the effect that technology is about to open gates either to Heaven or to Hell, that technology now presents man with some simple and decisive choice between immolation and utopia. Choices on earth seem likely to remain more confusing; utopias may sometimes be hard to tell from Brave New Worlds; evils may be not stark and obvious, but subtle, slow-acting, uncertain, and well distributed among all the available options.

*   *   *

In trying to get a large view of man's prospects, as seems to be appropriate for this consultation, we might for purposes of discussion consider five past and future economic stages, as listed in Table 1, below. The first of these we would call "pre-agricultural";

TABLE 1
Stages of Economic Change

Economic System	Annual Per Capita Product in (1965) Dollars	Leading Sectors Appeared
1. Pre-agricultural		1st 500,000 to
2. Pre-industrial	50–200	1,000,000 years
(or agricultural)		8th Millenium B.C.
(Industrial revolution)	(Transitional)	(Eighteenth century)
3. Industrial	500–2,000	Nineteenth century
(mass consumption)	(Transitional)	(Twentieth century)
4. Post-industrial	5,000–20,000	Twenty-first century
5. Almost post-economic	50,000–200,000	Twenty-second century

almost every society remained in this level until about 8000 B.C. At about that time "permanent" (i.e., systematic and sustained) agriculture is supposed to have started in the "fertile crescent" of the Middle East. We can think of this as the beginning of the "agricultural" or "pre-industrial" stage. One way to characterize this stage is to note that the annual per capita product of a society was very likely to fall between 50 and 200 (1965) U.S. dollars. The level was not very much above the pre-agricultural—but many more people were supported at this level. Given this base, it became possible to divert about 5 or 10 percent of the agricultural product to build and maintain cities, and thus to create a "civic" culture and "civilizations."

It has been argued that until the industrial revolution no society ever produced much more or less than $50 to $200 per capita per year for any extended period. During the late eighteenth century, the industrial revolution, which in some ways had begun much earlier, became an obvious facet of English society. Industrialization then spread rapidly in the West, more slowly elsewhere, and, of course, has yet to reach many societies. (We can think of an "industrial society" as having a per capita product of about ten times that of a pre-industrial, i.e., between $500 and $2,000.)

Today in the U.S., Western Europe, and Japan, industrial society has been gradually transformed into what sometimes is called a mass consumption society. We think of mass consumption as a transitional stage to what we will call the "post-industrial" society —much as the industrial revolution can be thought of as a transitional stage between pre-industrial and industrial stages.

This post-industrial stage can be characterized in many ways. We will define it here by the achievement of another increase of a factor of 10 or so over the industrial society in per capita product (to perhaps $5,000 to $20,000 per capita). By the year 2020, about 20 percent of the world's population should be living in such societies. We can of course expect many important qualitative changes to occur in these societies, in addition to those associated with the simple and direct effects of being richer. We can also envisage that if anything like current growth rates continue, at some point between the end of the twenty-first and the early twenty-second century we should see the beginnings of what might be thought of as the "almost post-economic" society, in which annual gross product per capita ranges from $50,000 to $200,000. By "almost post-economic" we mean that at this level of affluence a great many traditional economic issues will disappear or become minor, though, of course, others will remain and there will be some new ones.

Our task here is to project fifty years ahead. If current trends continue the world's nations might be distributed about as follows. Perhaps 5 percent—perhaps less—of the world's population should live in nations that might roughly (and in many ways inaccurately, by then) be termed pre-industrial. At the other extreme, 5 percent or so might be living in nations that are either beginning to show "post-economic" features, or are manifesting them clearly. In between, about 20 percent of the world's population might live in states that are clearly post-industrial, some even approaching or entering the "post-economic stage; about 10 percent might live in states which one could think of as mass-consumption societies (that is, in transition between industrial and post-industrial); another 10 percent of the world might live in what we might think of as mature industrial societies; and more than half the world might live in nations which we can regard as partially industrialized. Of this last group more than 90 percent (or about half the world's population) is likely to live in six large nations: China, India, Brazil, Pakistan, Indonesia, and Nigeria. Whether or not one or more of these six nations passes our rather arbitrary lower bound of "industrial" status, at $500 per capita, they will all have large cities and other enclaves which will certainly be at least industrial and parts of which might even be post-industrial. On the other hand, there will be many areas in these "large and partially industrialized" countries, particularly some or most rural areas, that for the most part will be in a pre-industrial, usually sixteenth or seventeenth century, condition, modified, however, by the addition of such industrial products as the electric light, transistorized radio and television, bulldozers, and even crop dusters and fertilizers. Such disparities of income and level of modernization may create very important stresses within as well as between countries.

We would expect to find almost all of Europe, including the Soviet Union, but possibly excluding some eastern European countries and some countries on the southern and eastern rim of Europe, such as Portugal, Spain and Ireland, in the post-industrial, and to some extent in the almost post-economic categories. However the excluded countries would presumably have reached at least the "mass-consumption" level. They would probably be joined by such Latin American nations as Argentina, Venezuela, Mexico, Colombia, etc.; a good deal of the Sinic culture area, outside China, such as Taiwan, South Korea, Philippines, Malaysia, etc., the Union of South Africa and possibly some of the Arab countries and Turkey (assuming there are no disruptive changes in these states). Much of Africa and some of China and Latin America might still be only partially industrialized—even if economic growth rates become quite high and population growth rates eventually decrease, but assuming both of these still stay within reasonable current expectations. In such a world there will be opportunities for creative and exciting lives for many; and orderly, decent, and remarkably full lives for the mass of people in the unprecedentedly affluent societies. Consider the outline below, and note that it is likely that while these conditions have improved significantly in France and in most highly developed nations, by the year 2020 no part of this account will be characteristic of any major society.

A Typical Life History of a Married Seventeenth-Century Frenchman

1. At birth, had only one living grandparent.
2. Had four brothers and sisters.
3. Lost one of his parents by the time he was an adult (15 years).
4. Lost two or three of his siblings by the time of his marriage (27 years).
5. Lived through two or three famines, three or four near famines, and two or three major epidemics (in addition to semi-permanent epidemics of whooping cough, scarlet fever, and diptheria).
6. Died at 52.

In addition to an enormous decrease in the actual amount of human suffering and tragedy, and of anxiety about the immediate prospects of life, security and making a living, there will be many positive things which man will be able to do. Both increased wealth and increased technology increase enormously the range of possibilities open to all—though they do not guarantee that the new possibilities will be put to good use.

*   *   *

While the developments listed below [Table 2] are, of course, by no means as spectacular as modern weapons technology, they raise very serious issues:

TABLE 2

Gradual and/or National Contamination or Degradation of the Environment

---

   a. radioactive debris from various peaceful nuclear uses.
   b. possible greenhouse or other effects from increased CO_2 in the atmosphere.
   c. waste heat.
   d. other special wastes.
   e. other wastes, debris, and just plain garbage.
   f. noise, ugliness, etc., associated with many modern activities.
   g. excessive urbanization.
   h. excessive overcrowding.
   i. excessive tourism.
   j. insecticides, fertilizers, growth "chemicals," food additives, etc.

These are primarily nuisances, rather than matters of potential life and death for hundreds of millions of people, but nuisances that might grow serious enough to threaten man's individual health or survival, or his general economy, comfort and/or happiness. In almost all cases technologically feasible cures exist, given the money to develop them. "Only money" is needed to control contamination and degradation of the environment. But this requirement may still put sharp limits on man's economic progress. In the past we had no need to worry about such problems; more immediate economic issues dominated. Traditional economic issues seem likely to be much less troublesome by the year 2020. While people in all walks of life often worry about the possibility of running out of natural resources, we would argue that for the year 2020 this possibility is usually grossly exaggerated—at least as the phrase "lack of resources" is usually understood. Such resources as water, fertile land, or minerals are not likely to set the crucial limits in the year 2020. On the other hand, we may run out of various kinds of available space—using this term in the most general sense—for various activities. This is true of many natural and laboratory situations. There are cases in which the limits on the growth of bacteria or other animal life in a crowded area are set not by any shortage of food, water, or air, but by a shortage of "lebensraum," or by the accumulation of waste products. Until the present century, by and large, man could afford to dump his waste, debris, products of combustion, and other consequences of his various activities into the rivers, lakes, oceans, and the atmosphere, or just leave them lying around. If the environment became unpleasant, men would either accustom themselves to the degraded condition or move away and start elsewhere. This is becoming less and less feasible in more and more places in the world. And it seems clear that by the year 2020 an incredible degree of control—at least by today's standards—will be needed, and that these controls may be very expensive or limiting in various ways.

Most of the points of Table 2 will be self-evident to the reader. We might note, however, that the waste heat problem, while not greatly publicized, may become crucially important. There are nuclear power plants and even commercial steam plants which give off enough heat to raise at least the local temperature of a river such as the Colorado by several degrees. As the number and capacity of power plants increase, the issue of heating these rivers can become quite important. Similarly, in large urban areas the temperature seems to be 5, 10, or more degrees higher than in the corresponding rural areas, largely but not completely because of the rejection of heat from man's urban activities. It seems likely that as urban life becomes more affluent, larger in area, possibly more dense and more dependent on energy-consuming (and therefore energy-rejecting) devices, this problem will increase in importance. (For example, the air conditioning for an apartment house rejects much more heat to the outside world than is removed from inside the house, because of inefficiency.)

There are other special activities, some of which are not widely publicized, which can cause serious problems. Some experts feel that the burning of gasoline in the upper atmosphere by high-altitude jets may trigger odd and possibly dangerous reactions in a zone of many delicately balanced processes and delicately stabilized situations.

Table 2 also reminds us that urbanization itself—that is, the loss of rural or wild areas and the like, and their replacement by city streets and buildings—is considered by many a serious loss, particularly if this urbanization swallows up especially attractive natural areas. Indeed, just having a lot of people around is annoying in all kinds of ways. Although people are usually gregarious and tend to congregate together, most human beings also seem to like and need space and privacy, at least part of the time.

"Privacy" is a complex issue. It can involve such things as

a. the right to idiosyncratic
   thoughts
   utterances
   values
   way of life
   style and manners
   methods of self-expression
b. isolation or protection from
   selected aspects of the physical environment
   selected aspects of the social environment
   many pressures and/or other intrusions by individuals, organized private groups and businesses, and political and governmental organizations
c. the right to
   withhold information
   make many family and personal decisions
   be oneself
d. enough elbow room
   to be unobserved occasionally
   for aesthetic purposes
   to get things done
   as a value in its own right

*   *   *

Finally we will mention an issue which has received much publicity and which we would argue has probably been exaggerated in discussion—at least as far as conditions of today are concerned—viz., the use of chemicals and other artificial additives at various places in the food production chain. Of course, while there may have been excessive apprehension about this issue in the past, conditions could clearly get a great deal worse, and probably would, if it were not for such campaigns of "exaggeration." The year 2020 may see not only the presence of harmful additives but a continuing sacrifice of taste and of other aesthetic qualities to increased economic efficiency. This has already happened in the United States where commercial fruits and vegetables are inferior in taste to those which once were available (and still are in parts of Europe). However, the decreasing importance of economic efficiency in the most affluent countries . . . lends some hope that such trends might be reversed in those countries while spreading to the rest of the world.

A number of developments remind us that the contamination or degradation of the environment need not be either gradual or local. It could be spectacular and/or multinational, or even planetary in scale.

Spectacular and/or Multinational Contamination or Degradation of the Environment

a. nuclear war.
b. nuclear testing.
c. bacteriological and chemical war or accident.
d. artificial moons.
e. projects West Ford, Storm Fury, Starfish, etc.
f. supersonic transportation (shock waves).
g. weather control.
h. big "geomorphological" projects.
i. million-ton tankers (*Torrey Canyon* was only 120,000 tons) and million-pound planes.
j. other enterprise or mechanism of "excessive" size.

a. and b. *Nuclear War, Nuclear Testing.* We can pass over these two items without comment.

c. *Bacteriological and Chemical War or Accident.* We need only note that bacteriological and chemical substances can cause contamination not only in war; very serious and widespread damage might also occur as the result of a laboratory accident or even the development of an organism capable of producing an uncontrollable epidemic.

d. *Artificial Moons.* The "artificial moon" is a concept of the

Pentagon's "Project Able." The idea was to orbit a giant reflector which would catch the sun's rays and bounce them back to a pre-determined point on earth—in the case of Project Able, upon the jungles of Vietnam at night, thereby transforming night into day. Astronomers made the most vociferous objections to the proposed project, claiming that it would hamper their observations. A commit-tee of the Space Science Board of the National Academy of Sciences undertook a study of the proposal, the results of which were released early in 1967. The committee reported that "it saw 'no scientific value in a satellite reflector system that is in any way commensurate with the costs and nuisance to science of such a system.'" Beyond the effects upon science itself, the committee discussed possible adverse effects an artificially prolonged day might have on those plants and animals (such as the grunion fish which spawns at the full moon) whose life cycles are regulated by the natural passing of day into night. On the strength of these objections, and possibly of others as well, the government has decided not to pursue the project at present.

e. *Projects West Ford, Storm Fury, Starfish, etc.* Project West Ford was the name given to a project at the Lincoln Laboratory of the Massachusetts Institute of Technology which placed in contin-uous orbit around the earth a belt consisting of some 480 million very fine dipoles, each about .001 inch in diameter (this is about ⅓ the diameter of the average human hair) and about 1.5 cm. long. There were two major objectives in this configuration of a "reflector" for radio communication: (1) it would be substantially indestructible by enemy action, and could therefore provide reliable communica-tion in a wartime environment, and (2) by providing a continuous belt which would appear stationary if placed in the equatorial plane in a circular orbit, it would have substantially eliminated certain difficult antenna tracking problems involved in other systems of communication by satellites in relatively low orbits. Much of the controversy surrounding this project came from those who feared the belt would interfere with optical and radio astronomical ob-servations by reflecting man-made radio signals or solar rays. Actually, no such interference is known to have occurred in any significant degree. Others feared that once in orbit, the dipoles would be irretrievable and might interfere with some future, as yet unknown scientific observation. For this latter reason the dipoles were so orbited as virtually to assure that they would return to earth within about three years. The major fear relating to this project, once the initial fears were assuaged, was that if it were successful, a proliferation of dipole belts would occur which, by their very number, might sometime interfere with scientific enterprise.

Starfish was the explosion of a 1.4 megaton nuclear device in the Van Allen belt, in 1963, in order to study the effects of what was expected to be—and probably was, we hope—a *temporary* disruption in the belt, which screens off significant portions of solar radiation. Fears were widely expressed, and to some extent remain, that such an experiment could cause a permanent change in weather or radio communications.

Storm Fury was a joint Weather Bureau and Navy Project Series designed to obtain scientific data on the structure and nature of hurricanes and tropical storms and to tame them if possible. Several experiments have been undertaken for this purpose including one which bombarded the central force of a storm with silver iodide crystals which, it was hoped, would set off self-destructive forces within the storm center. Other experiments attempted to alter the structure of the storm by seeding rings of clouds at some distance outside its "eye."

A great deal of controversy ensued after the announcement of these projects, mainly showing concern that man's interference might intensify rather than pacify a storm. The Storm Fury projects have therefore been limited to storms which are more than 36 hours "away" from any populated area. Consider not only the potential dangers of storm-altering capabilities, if they were to get into the hands of irresponsible persons or governments with malevolent intent, but the legal and moral dilemmas that will arise regularly for an agency that has the power to shift—or to refrain from shifting—the path of a storm. For example, what kind of trade-off between lives and property will be in effect? Once the pattern of storms has been disturbed by deliberate interference, can the agency ever escape responsibility by failing to act?

f. *Supersonic Transportation (Shock Waves).* A much debated subject today is the sonic boom produced by supersonic jet transports. Some expect this difficulty to be alleviated by more efficient aircraft design (improved lift/drag ratios) or special operating procedures, but it is still an open question—most experts believe it a very serious problem. Indeed many now believe that supersonic transportation will have to be restricted to routes over water or other uninhabited areas. But the shock waves created by jets may also be excessively disturbing to ships at sea or to other aircraft.

g. *Weather Control.* Weather control of course is not really a contamination issue but it does change the environment in significant ways, and while it helps some it may degrade the environment for others.

h. *Big "Geomorphological" Projects.* Similarly, any big project which significantly alters the shape of the terrain or the physical landscape can have innumerable unanticipated effects. For instance, in the Great Lakes region of the United States, the construction of a canal in 1932 is still having adverse effects. The canal, built between Lakes Ontario and Erie, inadvertently permitted the parasitic sea lamprey to enter all of the Great Lakes where it preyed upon other fish. Trout, blue pike, whitefish, and other fish species having all but disappeared from the lakes, another relatively useless species has proliferated to the extent that it has become a nuisance, congregating near shore, clogging water intakes, and often dying and littering the beaches. Only now is the situation being brought under control, and that only by chemical warfare against the lamprey.

Other examples of far-reaching effects of man's hand upon the terrain can be noted when dams are built, inundating previously dry land. One case of particular interest occurred in the tropical areas of the Surinam River. The problems encountered ranged from the creation of swampy shore zones, which could give rise to "the development of insect life and problems of public health," to an oxygen shortage in the waters caused by decaying plant matter and leading to the death of countless fish. Certain kinds of weeds may begin to grow in such an area or spread beyond control, playing havoc with navigation and the remaining fish life. The possible ecological problems of such a project can be enormous.

i. The oil tanker *Torrey Canyon,* with a gross displacement of 120,000 tons, recently caused well publicized difficulties in the English Channel. Today, 300,000-ton tankers are already being built, 500,000-ton tanker orders are being negotiated, and people have begun talk about million-ton tankers. Similarly, people are building planes of 500,000 pounds gross weight and soon we will be talking about million-pound planes. A crash by such a plane in a city would cause immense damage—as well as killing and injuring 500 to 1,000 passengers. Similarly, consider the consequences of a collision between two of these huge planes. In general, enterprises or devices of "excessively" large size are likely to be created in the future. Some will be successful and safe, but at times things will go wrong; the effects will undoubtedly be large and catastrophic.

* * *

We are far from suggesting that the processes of technological and economic development can—or should—be reversed. Yet few of us are likely to be able to feel the optimism of the men of the Enlightenment at the beginning of our age of progress. Many of the unpleasant possibilities mentioned in this paper are avoidable. Others are inherent in present and accelerating tendencies of contemporary society; the momentum of research, development, innovation, diffusion, investment, and growth, is almost surely not reversible by anything short of a holocaust. Man is developing enormous power to change his own environment—not only the outside world, but also his own physiological and intra-psychic situation. The prevailing humanist view, which we share, is that this is "progress"—it would be no more desirable than feasible to attempt to halt the process permanently, or to reverse it. Yet our very power over nature threatens to become itself a force of nature that may go out of control; the social framework of action could obscure and thwart not only the human objectives of all the striving for "achievement" and "advancement," but also the inarticulate or ideological reactions against the process. By the early twenty-first century, we shall have the technological and economic power to change the world radically, but probably not yet very much ability to restrain our strivings, let alone understand or control the results of the changes we will be making. But if we cannot learn not only to take full advantage of our increasing technological success, but also to cope with its dangerous responsibilities, we may only have thrown off one set of chains—nature-imposed—for another, in one sense man-made, but in a perhaps deeper sense, as Faust learned, also imposed by nature.

If there is any single lesson that emerges from the above, it seems to be this: while it would certainly be desirable and might even be helpful to have a better grasp of how social action may lead to unanticipated or unwanted results, it is not likely to be sufficient. Given man's vastly increased power over his internal and external environment, and, in particular, given the unprecedented opportunities for centralization of social control that follow from the economic and technological changes that have occurred and that are likely to occur with ever increasing impact, the effects of social policies—planned or haphazard—are likely to increase drastically, and the consequences of mistakes are likely to grow correspondingly disastrous. While all decisions are in a sense irrevocable, this existential fact must be appreciated increasingly as it becomes an ever more dominant aspect of all policy decisions.

Of course, it will be worthwhile to try to improve our understanding of future possibilities and the long-term consequences of alternative policies. But the problem is ultimately too difficult, and these efforts can never be entirely successful; almost the only safeguard that then remains is to try in general to moderate Faustian impulses to overpower the environment, and to try to limit both the centralization and the willingness to use accumulating political, economic, and technological power. It would be nice if this generation could somehow arrange matters so that the inescapable increase in regulation of human choices remains in the hands of people who will respect its disastrous potential and will not unnecessarily centralize it further. We must somehow preserve a concern for perpetuating (or devising) institutions that can both limit and protect the freedom of dangerous choices—not only for today's individuals and the pluralistic social groups that would want their views represented, but more important, for those who will follow us—those who in the future may experience their problems differently and would not want to find that we have already—unnecessarily and unwisely—foreclosed their choices and altered their natural and social world irretrievably.

Suggestions for Additional Reading

The environmental controversy, and the far more comprehensive (if imprecise) definition of ecology that has emerged, has virtually created a new field of historical inquiry. This situation is undergoing rapid change. Fortunately, two excellent studies provide a broad historical background for the modern debate. The European background, from ancient times to the eighteenth century, is covered in a splendid book by Clarence J. Glacken, *Traces on the Rhodian Shore* (Berkeley, 1967). The topic is then very tidily carried through American history by Roderick Nash, *Wilderness and the American Mind* (New Haven, 1967). Nash's book concentrates on the history of conservation, but includes significant materials for a general understanding of American thought and action on the environment. Readers ought also to investigate Leo Marx, *The Machine in the Garden: Technology and the Pastoral Ideal in America* (New York, 1964), Howard Mumford Jones, *O Strange New World* (New York, 1964), Hans Huth, *Nature and the American: Three Centuries of Changing Attitudes* (Berkeley, 1957), Stewart Udall, *The Quiet Crisis* (New York, 1963), and Arthur A. Ekirch, Jr., *Man and Nature in America* (New York, 1963).

Among numerous contemporary accounts before the Transcendentalists, the reader should examine the writings, in numerous editions, of William Bradford, *Of Plymouth Plantation, 1620–1647,* J. Hector St. John de Crèvecoeur, *Letters from an American Farmer* (1782), Timothy Dwight, *Travels in New England and New York* (1821–1822), Alexis de Tocqueville, *Democracy in America,* William Bartram, *The Travels of William Bartram,* Philip Freneau, *The Philosopher of the Forest* (1781–1782), Thomas Jefferson, *Notes on the State of Virginia,* and Gilbert Imlay, *A Topographical Description of the Western Territory of North America* (1792).

Among the Transcendentalists, Emerson's views on man and nature extend throughout most of his writings. See also Sherman Paul, *Emerson's Angle of Vision* (Cambridge, Mass., 1952). Nor can the reader omit the writings of Henry David Thoreau. See not only the famous *Walden,* but also *A Week on the Concord and Merrimack Rivers* and *The Journal of Henry David Thoreau,* all in various editions. For more general studies of the role of nature in Romantic thought, see A. O. Lovejoy's two articles, "On the Discrimination of Romanticisms" in *Essays in the History of Ideas* (New York, 1955), and "The Meaning of Romanticism for the Historian of Ideas," *Journal of the History of Ideas,* Vol. 2 (1941), pp. 257–278. And see the brief but excellent analysis in Stow Persons, *American Minds* (New York, 1958).

Westward expansion and Frederick Jackson Turner's frontier thesis have not been examined from an ecological perspective, but some enlightening views can be derived from R. A. Billington, *America's Frontier Heritage* (New York, 1966), Arthur K. Moore, *The Frontier Mind* (Lexington, Ky., 1957), and Henry Nash Smith, *Virgin Land: The American West as Symbol and Myth* (Cambridge, Mass., 1950). Otherwise, late nineteenth-century environmental history is largely the story of the conservation movement. Any understanding of this phase depends upon Samuel P. Hayes, *Conservation and the Gospel of Efficiency: The Progressive Conservation Movement, 1890–1920* (Cambridge, Mass., 1959). See also Elmo R. Richardson, *The Politics of Conservation* (Berkeley, 1962), John Ise, *Our National Park Policy: A Critical History* (Baltimore, 1961), and the excellent reader by Roderick Nash, *The American Environment: Readings in the History of Conservation* (Reading, Mass., 1968).

Contemporary to the beginnings of the conservation movement was George Perkins Marsh, who is served well by David Lowenthal's biography, *George Perkins Marsh* (New York, 1958). For John Muir, see Linnie Marsh Wolfe, *Son of the Wilderness* (New York, 1945), and Muir's own writings, including *The Story of My Boyhood and Youth* (Boston, 1913), *A Thousand Mile Walk to the Gulf* (Boston, 1916), *My First Summer in the Sierra* (Boston, 1911), and *The Life and Letters of John Muir,* edited by William F. Bade (Boston, 1923). For Gifford Pinchot, see biographies by M. Nelson McGeary (Princeton, 1960) and Martin L. Fousold (Syracuse, 1961). Pinchot's own views are expressed in *Breaking New Ground* (New York, 1967). On Theodore Roosevelt's views, see *Letters of Theodore Roosevelt* (Cambridge, Mass., 1951–1954), and *The Winning of the West* (New York, 1889). Also see Paul Russel Cutright, *Theodore Roosevelt: The Naturalist* (New York, 1956). There is also a wealth of material in various issues of the *Congressional Record,* the *Sierra Club Bulletin,* and popular journals of the day.

Two comprehensive collections of essays will aid the reader in understanding the exuberant and widespread interest in ecology that sprang forth in 1969–1970. See Paul Shepard and Daniel Mc-Kinley, *The Subversive Science: Essays Toward an Ecology of Man* (New York, 1969), and *The Fitness of Man's Environment* (Washington, D.C., 1968). Both introduce the lay reader into some of the scientific aspects of the debate, as well as a variety of humanistic interpretations. There are also three volumes of essays edited by William R. Ewald, Jr., broadly based upon an urban planning perspective: *Environment and Change: The Next Fifty Years* (Bloomington, Ind., 1968), *Environment and Policy: The Next Fifty Years* (Bloomington, Ind., 1968), *Environment for Man: The Next Fifty Years* (Bloomington, Ind., 1967). Besides the writings of Carson, Sears, Ehrlich, and Eiseley that are represented by selections in this volume, the debate of 1969–1970 was influenced by a number of widely distributed paperback books for public consumption, mostly in anticipation of Earth Day on April 22, 1970. Perhaps most important was *The Environmental Handbook,* edited by Garrett de Bell (New York, 1970). Close on its heels was the volume by the Reinows excerpted in this reader, which should be read in its entirety. Other paperbacks were *Ecotactics,* edited by John G. Mitchell (New York, 1970), Wesley Marx, *The Frail Ocean* (New York, 1967), *Voices for the Wilderness,* edited by William Schwartz (New York, 1969), and *Earth Day—The Beginning* (New York, 1970). Nor should the press and journals be omitted. Almost every day *The New York Times* contains articles on all aspects of ecology; see also its weekly "News of the Week in Review" and *Magazine.* Monthly magazines, such as the *Atlantic Monthly* and *Harper's,* contained important articles during 1969 and 1970. *The New Yorker,* which originally published Rachel Carson's *Silent Spring,* continued its coverage, and the weekly *Saturday Review* instituted an Environment section, as did *Time* magazine.

Numerous other studies of diverse sorts can inform the student of the controversy. The variety of approaches taken, disciplines involved, and positions established is remarkable. See, for example, Ritchie Calder, *After the Seventh Day: The World Man Created* (New York, 1961), Marston Bates, *The Nature of Natural History* (New York, 1950, rev. ed., 1961), Philip Wagner, *The Human Use of the Earth* (New York, 1960), Hans Jonas, *The Phenomenon of Life: Toward a*

Philosophical Biology (New York, 1966), R. Buckminster Fuller, *Utopia, or Oblivion: The Prospects for Humanity* (New York, 1969), or his *An Operating Manual for Spaceship Earth* (New York, 1969), Joseph Wood Krutch, *The Great Chain of Life* (Boston, 1956), Harold and Margaret Sprout, *The Ecological Perspective on Human Affairs, with Special Reference to International Politics* (Princeton, 1965), Barry Commoner, *Science and Survival* (New York, 1966), Desmond Morris, *The Naked Ape* (New York, 1969), Robert Ardrey, *The Territorial Imperative* (New York, 1968), Ian McHarg, *Design with Nature* (n.p., 1969), Peter Farb, *Ecology* (New York, 1963), and the quarterly, *Whole Earth Catalog* (Menlo Park, Calif., 1969–1971). And finally, the writings of Rene Jules Dubos cannot be ignored, particularly his *Man Adapting* (New York, 1965), and his 1969 Pulitzer Prize volume, *So Human an Animal* (New York, 1968).

2 3 4 5 6 7 8 9 10